AF754015

Little Hopper

Black Widow

Pink Wolf

Blue Max

Tadpole

Hooked on Flies

Confessions of a Pattern Inventor

Hooked on Flies

Confessions of a Pattern Inventor

WILLIAM C. BLACK

Copyright© 1980 by William C. Black
All rights reserved
Library of Congress Cataloging in Publication Data

Black, William C 1931-
Hooked on flies.

Includes index.
1. Fly tying. 2. Flies, Artificial.
I. Title.
SH451.B6 799.1'2 80-16611
ISBN 0-87691-312-5

Published by Winchester Press, Inc.
1421 South Sheridan
Tulsa, Oklahoma 74112

Printed in the United States of America

Book Design by Nancy Steinmeyer

1 2 3 4 5 6 7 8 9 — 86 85 84 83 82 81 80

Introduction

It has been my enviable fortune to have known a number of very good fly tyers, both amateur and professional, over nearly half a century, and in several instances to have spent long hours sitting beside them as they created exquisitely beautiful, marvelously wrought trout or salmon flies out of the chaos of fur and feathers and tinsel that littered their tables and shelves. Lee Wulff . . . George Harvey . . . Polly Rosborough . . . Larry Madison . . . Ernest Schwiebert . . . Jim Deren . . . Olaf Olsen on the Laerdal . . . "Big Jim" Leisenring . . . Walt and Winnie Dette . . . Harry and Elsie Darbee . . . Stan Cooper . . . Elizabeth Gregg, the thrifty Scots lassie who taught me to tie "in the hand" and thus save the expense of a vise . . . and a host of less renowned but skillful craftsmen.

You might think some of the talent of these men and women would have rubbed off on me. But although I learned something from each of them and a great deal from some of them, fly tying is one of a vast curriculum of activities for which I have minimal aptitude. Since my trout-fly standards are considerably higher than my fly-tying ability, I buy most of my flies, or cadge them from kind-hearted fly-tying comrades, tying only a few non-standard and otherwise unavailable patterns.

One of the amateur tyers I *haven't* met in person is Dr. Bill Black, but I had so much pleasure (and some profit, fishing-wise, I hope) in reading his book that I jumped at the opportunity to write this introduction, like a rainbow trout jumping for one of the doctor's Golden Adams. Like a lot of amateur tyers and even some pros, and like Dr. Black whose inventions and modifications are the subject of this book, I've spent a good bit of time at the tying desk trying to put together, out of inspiration, intuition, and some thread and feathers and dubbing, the Ultimate Trout Fly, the 12 or 14 on a scale of 1 to 10, the be-all and end-all, the absolutely irresistible fly (although, come to think of it, we already have that kissing cousin to the Rat-Faced McDougal, but with a capital "I").

Alas, there is no such Ultimate. And how drab life would be if there were! No more eye-delighting fly-boxes crammed with a hundred gaudy patterns (only five or six of the dullest of which are reliable fish-getters). No more the frustrating kid-in-a-candy-store ritual of pawing through the trays of the Darbees' and Dettes' display counters, unable to afford a tenth of the goodies one would like to transfer to his own inventory. No more the joy of choosing a dozen new salmon flies tied with cunningly married exotic feathers and gleaming gold ribbing that rejoice the angler's eye and don't seem to bother the salmon much. It would be like a world in which all the people were mirror images in face and figure, and all the trout were of identical size and weight and coloration . . . ye gods!

But there *are* effective new patterns each generation, developed by bored or creative or psychotic tyers, and in a lifetime a gifted man or woman might possibly (but by no means probably) invent a new or improved trout fly that endured past his or her own span of years. Theodore Gordon did it. Roy Steenrod did it. Lee Wulff (whose span, happily, goes on apace) did it. Vince Marinaro did it. Whoever invented that abomination, the Bomber, did it. Maybe Dr. Black has done it. We won't know for a number of years.

What the doctor *has* done is to tie and test and discard a great number of new or modified patterns and to salvage a handful of innovations that proved to be outstandingly effective over an extended period of time and under a variety of conditions.

And although most of these were developed and tested on Western waters, they're almost all of them impressionistic rather than specific, and should be just as attractive to Midwestern or Eastern trout. But perhaps the most important justification for this book is that it may encourage many of its fly-fishing readers to experiment with new colors and materials and tying techniques, in as cautious or as reckless a way as suits their temperaments, and to keep careful records of which experiments, if any, are successful.

What Dr. Black points out in these pages, in gist, is that fly tying has its own dynamic—with room for the strict traditionalist who may reject terrestrials (except for the Cow-dung) and no-hackle patterns (except for the Hare's Ear) as well as for the rip-roaring innovator and iconoclast. A blessing on both their houses!

But I'm keeping you from the doctor's text. I think you'll find it as interesting as I did.

Ed Zern

Contents

CHAPTER 1

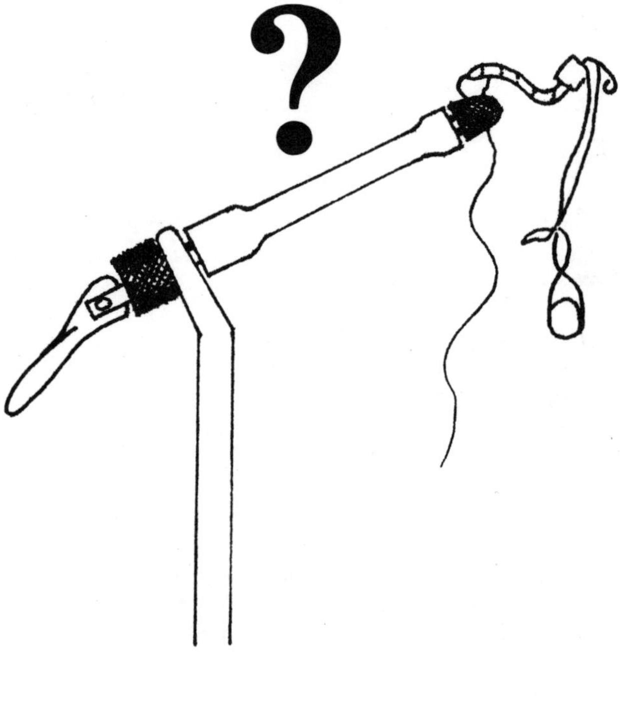

Why Bother?

The opening chapter in a fishing book is the most difficult to write; even the title is a problem. The commonest theme utilizes terms such as "early," "first," and "beginning," genesis words of a sort, as if something fairly remarkable had happened to or about the author. In my case it isn't so, and since the book has to do with making up your own patterns, I decided to get right to the meat of things by asking a head-on question: Why bother? And there's an implied corollary: What's wrong with the old familiar flies? So taking these in reverse order, of course there isn't anything wrong with the venerable favorites. If there were, they wouldn't be standards today. Beloved patterns such as the Gold Ribbed Hare's Ear go back at least a century and don't require advertising campaigns to maintain their place in the fly fishing public eye; there is no need for radio jingles or TV commercials. Instead, their reputations are based on years of dependable performance. There could be no sounder recommendation.

Getting back to the basic question, I've found that fooling around with fly patterns is just enormously entertaining. To begin with, fly tying is truly a hobby within a hobby, a special facet of fly fishing, and it's clear that many anglers simply enjoy the handicraft aspect of making a small thing, such as an artificial fly, nicely. There's a sort of neat perfection that's involved. Add to this the creative aspect of expressing your own ideas in one or another experimental pattern and the process becomes doubly satisfying. The word "experimental" brings up still another side, since you inevitably begin by wondering what the trout will think of an unusual tying trick and you know that you'll soon be able to actually find out. This is surely appealing, for the mechanism of asking a question that one can later answer is the basis of all research, scientific and otherwise.

Finally, there's the fact that the trout can hardly have seen anything quite like your homemade patterns before. Regardless of whether they are interesting to the fish or downright offensive, your flies should at least be new and novel! Now, this observation may have no significance whatever, but it pleases me to think that it might. On several occasions bass fishermen have told me about particular lures that were dynamite when new on the market only to lose some of their effectiveness gradually over a span of seasons.

In summary, it would appear that pattern invention or modification is a natural extension of fly tying per se. It is a fascinating pastime for the restless angler, the inquisitive fly fisher who likes to fuss with his tackle and ask questions about observations he makes on the stream, the person who is not satisfied to stick with the status quo.

But is there really such a thing as a totally new pattern? So many tyers have been working at their trade or hobby for so many years that "it" should all have been tried before. Surely every conceivable combination of fur, silk, and feathers must have come out of someone's vise at one time or another, whether or not formally named and recognized. Well, there's no doubt that the traditional patterns added to an array of newer ones present an impressive spectrum of variability, but there is still a basic or central theme insofar as types of construction materials, their colors, and the ways in which they are put together. I've found that if you'll make an effort to get out of this groove, it's not at all difficult to come up with some innovative if not always successful schemes. As you'll see, I've done this in the area of dry-fly wing color with some resultant creations that might politely be described as "strange." Just as interesting is an increasing number of potentially useful synthetic materials with which one can improvise. Many of these have nothing whatever to do with fly tying in terms of the purposes for which they were designed, and there's about as much room for explanation as the limits of your imagination will allow.

Having gotten this far through the preamble, many readers might like to make a pointed inquiry of this nature: What makes this guy, whom I've never heard of (probably), think he can develop trout patterns off the top of his head that have anything special to offer?

THE COCKROACH. . .

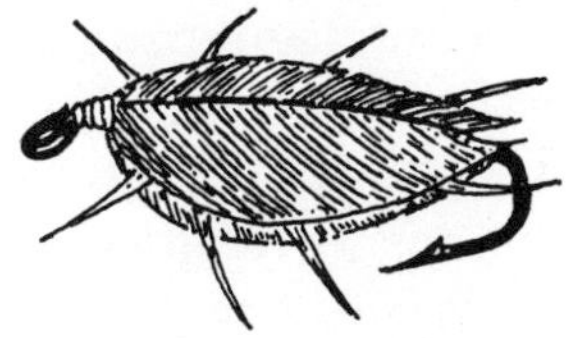

—(DON'T BOTHER)

I believe that I can prove I'm not an egomaniac by admitting that while I began manipulating patterns during high-school days some thirty years ago, my real talent is in developing flies that flop! Either that or ones that are no better than the standard patterns we have all used for years. This is where I excel. In fact, looking back, the ratio of failures to successes has been about twenty to one, and I'm haunted by countless unfortunate efforts like the cunning (I thought) Cockroach. To prove the point further, I've described several other spectacular disappointments in the final chapter, placed there for fear of discouraging the reader. So I think I have reasonable control of my self-image; after all, if you try enough ideas, sooner or later you're bound to stumble onto something that works just by sheer chance.

Another thing: I'm a lazy tyer. Honestly, I average only about 300 flies per season, wet and dry, and that hardly qualifies me as an expert. It would be a long time between meals for a professional with this kind of output. Frankly, I'd rather take a beating than have to sit down and tie a dozen flies in the same pattern; that's too much like work. Good tyers set out appropriate materials on the bench beforehand in the interests of efficiency and cost accounting, ready for mass production, whereas I'm an impulse tyer who finds the repeated construction of standard patterns dull and tedious. There's nothing to pursue beyond technical perfection. And I'm neither a fast nor a fancy tyer. I can't turn out more than half a dozen flies in an hour no matter how hard I try, and I'd rather not try. On top of this, I don't use a bobbin or a whip finisher and know only two tying knots. We'd better go on before I get depressed.

Do the ten patterns described in this book have anything to offer? That depends on two factors. First, I must interest you in trying some of them. This was the obvious challenge I set for myself in writing the book. Second, once on the leader, my flies will need to take trout for you, and I'm confident that they will. Ultimately, of course, you must be the judge.

The principal construction or design theme behind all of my labors of the vise is *impressionism.* I had absolutely nothing to do with originating the concept of impressionistic flies, but I'm a zealous disciple of impressionism. In brief, a pattern of this type is supposed to suggest or give an impression of a whole group of food forms of one sort or another. Rather than slavishly copying a particular type of adult insect bobbing in the surface film right down to genus and species, an impressionistic dry fly will suggest any member of a much larger and more diverse group of floating insects. The same principle can be applied to terrestrial bugs, submerged larvae (nymphs), and even minnows (streamers). As I pointed out in a book I wrote several years ago,

Flyfishing the Rockies, there is tremendous efficiency in this approach, since a single impressionistic pattern can cover a great many mimicry needs. Thus an angler doesn't need to rummage through bulging pockets for some special, compulsively perfect, hatch-matching imitation for each of the myriad of insects he may encounter. Impressionism is the simple way to go.

In the course of explaining how and why I developed each fly, I'll naturally be covering a good deal of water. But you'll find that few of the experiences I mention are typical "fish stories," exciting and dramatic tales in which I catch lots of trout or a very big trout or, better still, lots of big ones. Magazine writers use this technique to good effect, but the problem is I'd have to play the same tune throughout, chapter after chapter, success story after success story, and this would get old. Sooner or later my credibility would have to suffer. I also suspect that most readers are already experienced, perceptive anglers who realize that one stunning performance doesn't necessarily mean the fly will produce in another locale at a different time and under a new set of circumstances (assuming they believe the tale in the first place). The same applies to mentioning large numbers of streams and rivers, particularly the more famous trout waters, where one or another pattern has proved to be a winner. This is really a form of name-dropping involving supposed excellence by association. The implication is that the author must be knowledgeable since he writes about name places, and, further, that trout living in such celebrated waters would likely shun any but the best of patterns. This is a sort of "Madison Avenue" jive (a pun) I'll try not to inflict.

Another trap of triteness I've attempted to avoid is the overused set of adjectives that are applied to effective flies. Aren't you tired of hearing about patterns that are deadly, killing, or devastating? I'm perhaps unduly sensitive, but I'm sure mature, sensible fly fishers aren't out for vengeance when they go astream. They aren't preoccupied with dealing death to or otherwise devastating as many fish as possible; it isn't a war, after all, nor are the trout enemies. If we must count coup, what greater victory than magnanimously granting the captive a new lease on life!

I've assumed that almost all readers will also be fly tyers, since it seemed unlikely that non-tyers would be interested in a book like this one. Accordingly, this isn't a basic how-to guide. However, if there are a few anglers who buy all their flies and have somehow gotten into the wrong pew, let me urge them to take up fly tying just as soon as possible. It's not that hard. One needn't be an accomplished surgeon in order to learn quickly; indeed, it's better if you're not, in my experience. Nor is fly tying a suitable arena for the male chauvinist, if you see what I mean. (The same is true for fly fishing generally, although I have used fly fisher*man* throughout when referring to any angler; fly fisher*person* is just more than I can stomach.) As to age, I've worked with sixth-graders who progressed amazingly. Especially when it comes to wet flies, the younger tyer's creations tend to have a sort of devil-may-care look of reckless abandon that trout seem to trust. Either that or a whole series of kids I've fished with over the years have been awfully lucky! There are at least a dozen good books on tying basics available for the beginner today, plus all sorts of small local classes that can help with the rough spots.

My motive in writing this small book was not to try to establish any of my flies among the standards. That would be like tilting at windmills. Instead, I wanted to interest other anglers in pattern innovation or modification because it has been so much fun for me. Rather than accepting my patterns as finished products, I hope they'll be tampered with and improved upon. There's plenty of room, no doubt about it. I'd like to think that my artificials can provide sparks for ideas of your own. I believe that fly fishermen are experimenters by nature, so I hope you'll join in the quiet excitement that goes with pattern invention.

CHAPTER 2

The Spanish Fly

Any luck? That's the traditional greeting when anglers meet along the stream; it's almost a password. And if the reply should be of a positive nature, another query quite naturally follows: What did you catch them on? Or what are you using? If the other party is a bait or lure fisherman and if I have anything to report (many times not), he is usually satisfied with a very basic reply, i.e., "Flies." When another fly fisherman comes along, though, it's another matter. I can tell him that it's my own pattern or something I made up, but this is neither an informative nor particularly interesting communication. In fact, it's embarrassing, since it sounds as if I'm onto something hot I'd rather not share. Understandably, the other anglers seldom pursue the conversation further; they usually shrug and go on their way. The alternative involves naming the pattern as if they *should* know what I'm talking about, but how would you react to a silly answer such as "I'm getting them on a No. 16 Spanish Fly"? So there's no easy way out, nor are made-up patterns simple to explain at the streamside.

Oddly enough, the Spanish Fly is best understood in the context of the apparently unrelated Royal Coachman. You'll see why as we go along.

Sophisticated modern anglers seldom mention the ancient and honorable Royal Coachman. Perhaps there isn't that much to say about a pattern that goes back through so many generations; and I've thought too that the Royal's fine name and flashy appearance give it an air of superficiality, a sort of commercial image, that puts some of us off. Besides, this is an era when photographically exact reproductions of natural insects are very much in vogue, and the Royal is the very antithesis of this kind of hatch-matching artificial. Indeed, it's virtually a crimson-bellied anachronism. Yet I've always liked this honky-tonk fly. Without getting maudlin about the whole thing, it was an important part of my piscatorial boyhood. The Royal is very much an old friend, and one that has done a lot of fishy favors for me. Among these good deeds, it fathered the Spanish Fly.

It all came about because of two different problems I had with the Royal Coachman presented as a dry fly. In *Flyfishing the Rockies,* I described a simple mathematical observation you can make on the stream that I call the "strike:hook ratio," meaning the ratio of rises to hooked trout. Naturally the observed ratio varies with time and place and also with different patterns. Now in the case of the Coachman there were often days when my ratio would be quite high—say, five rises for every hooked fish. It seemed that the risers would spit out the fly or just sideswipe it at the last instant no matter how alert I tried to be and quick in setting the hook. Of course, for a fly fisher each rise is a pinpoint quantum of excitement, but still, a high strike:hook ratio adds up to frustration, and I began to search for remedies. I found that the ratio could sometimes be lowered by simply changing to a smaller hook, say, from a No. 14 to a No. 16. However, this was only sometimes. I had noted that two other popular artificials, the Rio Grande King and Black Gnat (male), also suffered high ratios from time to time, and it occurred to me that the three flies had one thing in common. Each displayed marked contrasts in color or shade—a white wing versus a darker hackle and body. I wondered if perhaps the trout "read" this sort of stark contrast as a last-second warning to the effect that they were about to swallow something phony, just in time to swerve away. At least it was a thought that could be tested, and it seemed easiest to begin by toning down the bright white wings somehow. Working with either calf tail or hackle tips, I dyed the white wings to an off color in either pearl gray or light beige, much as one might tone paint. This killed the contrast all right, but it did nothing good for the fly's performance. The dingy wings made the tinted Royal harder to see on the water and, if anything, less attractive to the trout. Later I tried other colors, lots of them—in fact, a whole rainbow's worth over a period of years. It took about eight seasons to finally prove to my satisfaction that *medium-gold* wings could, at times, improve the pattern. You can see that this kind of experimentation is very time-consuming, since a single angler can test only one fly at a time in one place and for a limited number of hours during each season. (I have a family to support.) From time to time friends help me try out my pattern ideas, but I've no right to usurp their coveted angling hours with my innumerable experiments, many of which have been less than fruitful, as explained. Thus, it can take several years for a pattern to demonstrate convincingly its true lack of value, for many times a fly will be just mediocre rather than blatantly bad.

As you've probably guessed, my game plan involved fishing whatever wing color I was trying against a regular Royal in the face of a high strike:hook ratio, and my gold-winged version almost always knocked the ratio down for me. It goes without saying that the ratio is a secondary consideration—you've got to get them to rise in the first place before getting out your pocket calculator. However, I found that the trout would generally come up for the fly with equal frequency whether it was white- or gold-winged, so the ratio really did matter. And the gold wings provided another quite unexpected bonus. Under certain water and lighting conditions they made the fly surprisingly visible, even more so than a regular white-winged Royal. To be specific, when there is severe surface glare mixed with tiny floating flecks of white foam, the usually obvious Coachman can practically disappear. Its wings and the foam flecks are very much alike against this difficult background, but interestingly, the gold just doesn't fit in; the color is different. As a consequence, your eye tends to pick up the gold-winged version rather easily.

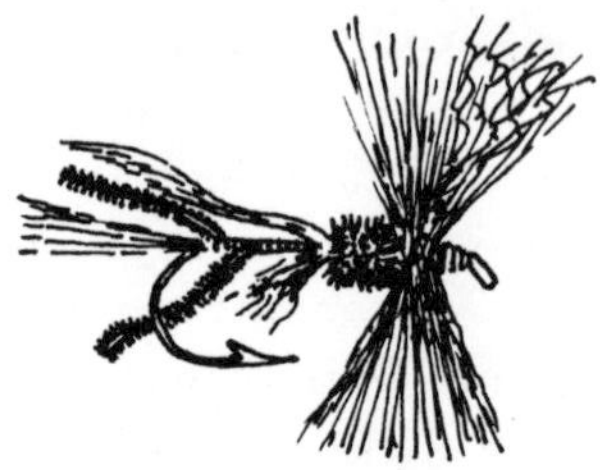

My second complaint about the Royal Coachman had to do with its fragile rear end, especially the rear cuff of peacock herl. The darn stuff would invariably break just when the trout were taking best. Then the red silk would promptly fray until the once neatly uniformed Coachman became a disheveled mess of trailing floss and spikes of herl. It happened that at the time I was quite fond of the Blue Variant, an uncomplicated wingless pattern consisting of blue dun hackle, a gold tinsel body, and hackle fiber tail. The tinsel body was virtually indestructible, so I thought about substituting gold tinsel for the triple-banded herl/floss/herl body of the Royal. Besides beefing up the fly's tender belly, the tinsel would blend nicely with both wings and hackle. On the other hand, I was reluctant to drop the peacock altogether, for it was an integral part of an awfully successful pattern. Since it was almost always the rear cuff that broke, I ended up with a compromise by retaining the forward, more protected muff just behind the wings.

When it comes to tails for dry flies, I'm afraid I show a singular lack of imagination in all of my patterns. To my way of thinking the tail has three contributions to make: It adds to the fly's silhouette, it should help with flotation, and it provides weight in the rear so that hair-winged flies won't tip frontward and float face down. For these tasks, natural grayish-brown deer body hair is ideal. I don't see the drab color as a drawback, and a reasonably thick, slightly stiff hair tail puts plenty of ballast back there to counterbalance the wings. Meanwhile the buoyant fibers trap tiny air bubbles between individual hairs to yield extra flotation. Beyond this, the light butt ends of the hair tied down beneath the tinsel give the body fullness.

So this is why the Spanish Fly is the offspring of the Royal Coachman, and I've also told you that it's both visible and durable. What you really want to know, though, is where and when and why it catches fish. To begin with, it is quite evident that, like the Royal Coachman, this pattern doesn't match anything from the real world of insects. Instead, it would be classified as a nonimitative *attractor* fly. Now, there's no particular shame in this. An attractor pattern such as the Spanish Fly ought to have a broad spectrum of uses, or more properly, of trout acceptability, and its effectiveness shouldn't be limited to times of special insect activity such as hatches. Therefore, the answer is the same as it would be for the Royal Coachman; you use the Spanish Fly at any time. Similarly, since there is no particular hatch matching involved, the pattern should do the job throughout the season and in a wide range of hook sizes too. I believe that the above profile does describe the Spanish Fly accurately; it simply has general appeal. This much-adulterated Coachman has earned a reputation (with me at least) for smoothly consistent productivity from spring through fall in trout waters of practically every type. I work most with hook sizes 14 and 16, although I've had success with butterfly-caliber 8s and midgelike 18s too. My records don't show any differences in species response; I do suspect that the Spanish Fly has caught me more browns overall than any other trout.

Trout are usually pretty nice about coming up for this fly whether or not there are naturals about. However, there is one observation that causes me to try a Spanish Fly ahead of other patterns, and this is the presence of yellowish-beige insects, frequently caddis species, cavorting over the water. In this situation the Spanish Fly has batted almost 1,000 so far.

It's time to go back over the fly's various pieces of anatomy in a little more detail. As to hackle, I've tried both dark brown (Coachman brown, as it's often called) and ginger and have come to prefer the ordinary medium shade that most anglers like for the Royal Coachman, in other words, red game. This is one of the least expensive and most easily available quality hackles on the market. I've used three wing-construction materials for the Spanish Fly: hackle tips, calf's hair, and polypropylene yarn. (Like almost everyone else, I can't abide quills, for they are fragile, are hard to tie and balance, and add nothing to a fly's floating qualities.) On some No. 16 hooks and all No. 18s I work with either polypro or hackle tips because of their delicacy, but I'm really a hair man at heart. Hair wings make a very real contribution to a fly's visibility and buoyancy. They make a clear and positive statement in the overall composition of a pattern, too, rather than being an add-on bit of trim or decorative afterthought. And don't get the idea that a hair-winged dry fly has to be coarse and crude. The length and thickness of the wings can be tailored to your own specifications. There's no need to anchor a huge hank of hair to the hook, billowing over the rest of the fly like a cumulus cloud; hair wings can be dainty if you wish. Polypro yarn comes in almost any color, including gold, and I dye my own hair wings. Here I'd like to make a deal with you: There are four other patterns with dyed wings to talk about, and it makes sense to avoid repetition by deferring a discussion of the particulars as to how it's done until the next chapter. OK?

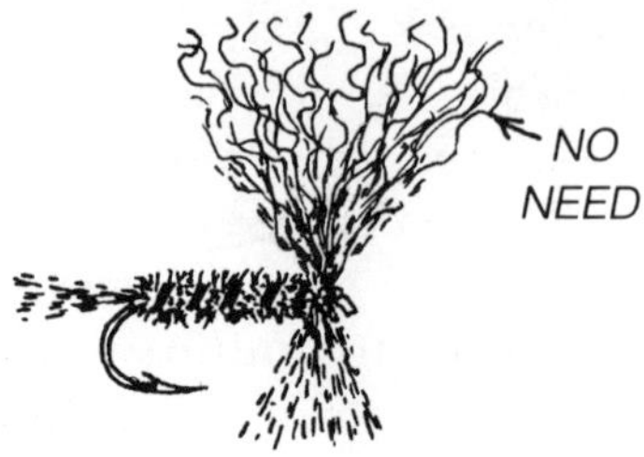

The way in which I put the body and tail together is a bit unusual. I begin by tying the tinsel (fine-caliber, oval with thread core) onto the top of the shaft. Then I tie in about twenty fibers of deer body hair clipped from the skin in a bunch at the hook's bend after aligning the naturally tapered ends. The tail should be about as long in relationship to the hook's gap (distance between point and shaft) as shown. A longer tail makes the body appear unduly extended. Allow the butt ends of the hair to splay about the shaft as they will, and back-wrap toward the eye, tying off and trimming just past mid-shaft. The tinsel has been sticking out to one side all the while, and now it gets wrapped on in tight spirals, tied off, and trimmed.

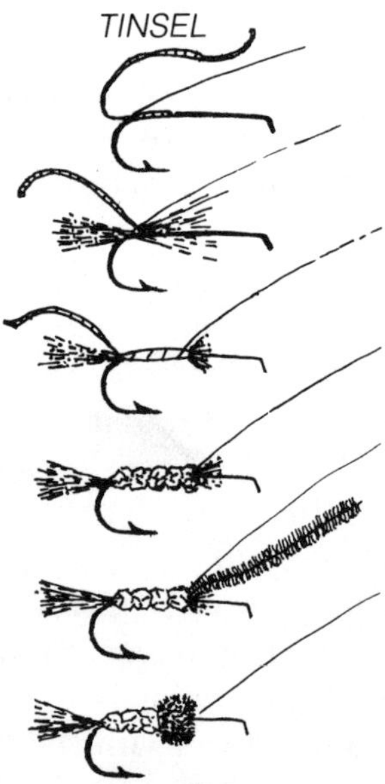

I put the tinsel end under the tail to avoid a tie-in lump and then wrap it over the tail butt, because the buried hair ends give the body fullness and buoyancy. Next I tie in a wide glossy peacock herl right on top of the tinsel tie-off and wrap on firmly, working back to mid-shaft and then forward again before tying off and trimming. Note that the herl goes on top of the tail end; this fat base makes it stand out as a distinct cuff when wrapped in on edge. I lacquer the tinsel lightly, but not the herl; cement would mat the glossy fibers into a lifeless dark-green band. All of this takes just a minute or so, and the resultant body is extremely durable.

Although I promised not to drop famous place names, I do want to get in one quick tale about the upper Green River in Wyoming, since it involves a very left-handed sort of compliment for the Spanish Fly. Those of you who have worked the headwaters of the Green know that it is a haven for the Rocky Mountain whitefish. So willing are the foolish whitefish that it's hard not to catch them. If the mouths of these ravenous creatures were larger, I believe they'd attempt to swallow almost anything. I've seen them try for cigarette filters. In all honesty, the real challenge is beating the whitefish off your hook long enough to give the trout a chance. It's no exaggeration. If you spot one close to the surface that manages to gum your fly (they're toothless) without getting hooked, it's quite likely it will charge again on the very next cast like an enraged bull. You know it has to be the same fish, too, when you remove the fly and find a second fresh mouth wound. This happens not infrequently, and the catch ratio of whitefish to trout can easily exceed 20 to 1. Since there's not a whole lot of glory in catching these scaly, bird-beaked dummies, I've tried to learn how to avoid them. One ploy is to fish the faster water at the heads of currents as opposed to pool tails where the whitefish school in veritable armadas. (I've seen rises at dusk in the slow glides so thick that the entire surface boiled gently and whispered like a grove of aspens in the breeze.) Another gambit is to use spinning lures or huge flies. However, being stubborn by

nature, I fought the whitefish war for the better part of a week with conventional wet and dry flies. When it was all over, I'd failed, or so I thought, since for every trout I'd caught there had been eight simple whitefish (every one taken against my will, so to speak). Therefore, it was a surprise when I reviewed my log book some weeks later and found that most of the trout were caught on a Spanish Fly! I even caught more trout than you-know-what on the pattern. Now this is a silly story, since the same thing might not happen again and few readers will have a whitefish problem anyway. Still, I thought it was sufficiently unusual to pass along.

I get asked fairly frequently how the Spanish Fly got its name. One wag suspected that I might dip the finished artificials in dry-fly oil spiked with cantharis (the drug name), thereby exciting the trout beyond the angler's wildest dreams. Well, the fly isn't *that* good, and besides, it wouldn't be sporting. Actually, I put the finishing touches on the pattern while fishing the pretty streams of northern New Mexico up in Rio Arriba County. These are the watersheds of historic landgrants and share, together with surrounding peaks, passes, and tiny communities, lovely lilting Spanish names such as Tierra Amarilla ("Yellow Earth"). Since the trout were in fact eating my new pattern alive when the name came to me, I thought, Why not? And that's all there is to tell.

CHAPTER 3

The Golden Adams

If there were such a thing as the All-American Fly, it would be the Adams. It's a great old name in American history, and, more important, the Adams has a well-deserved reputation for dependability and solidarity in the best tradition of Uncle Sam. It's an honest, homely pattern rather than a flashy fly, and fooling around with the Adams would be a little like scrambling the stars and stripes in the flag. Inevitably, this pattern reminds me of John Atherton's fine book *The Fly and the Fish*, published over twenty-five years ago. This book is the fly tyer's doctrine of "impressionism," for Mr. Atherton was a highly practiced angler and tyer, as you'd expect, but he was also a professional artist, and he applied the concepts of the impressionistic school of art to the construction of his patterns. It's a fair guess that the Adams was probably his favorite.

It would be dishonest if I claimed to possess an in-depth understanding of the finer physical, philosophical, and psychological points of impressionism—the artistic "innards," so to speak. However, I do comprehend that *light* and consequently *color* are vital components which are employed to convey an illusion of life and vitality in a scene. This doesn't imply a smear of vivid hues so much as the thoughtful application of colors, perhaps in small quantity and in carefully planned combinations. For instance, an impressionist painter wishing to make a green tree come alive might cluster tiny flecks of blue and yellow to represent foliage. At a distance the eye automatically mixes these pigments, yielding an impression of a green that is vibrant and living. If the blue and yellow were premixed on the palette, the effect would not be the same at all. Another application of this same scheme involves the use of bright or light colors against a more somber background to suggest reflected light, and hence reality. Quoting from Chapter 8 of Atherton's book: "If you will look closely at a live dun . . . you will observe that his coloring is impressionistic. It is built up of many tiny variations of tone such as we find in the paintings of Renoir, Monet and others. And as they studied the form which reflected or absorbed light and thus took on certain color qualities of its surroundings, they were dealing in life, not death. Anglers should do the same." Now, the Adams was the leader of a compact group of six dry flies that were practically the only patterns Mr. Atherton used on the difficult Eastern waters near his home. Remember that a single impressionistic fly yields a lot of mimicry mileage by suggesting a variety of insects rather than just one species.

How does the Adams fit the impressionistic mold? It's a beautifully simple and yet subtle construction recipe based on mixing two feathers of different color in the hackle. I can best explain by borrowing another quotation from Ray Bergman's *Trout,* this time pertaining directly to the Adams: "This mixing of hackles is very effective; it promotes life in the fly and light showing through the combinations gives an iridescent effect to which trout seem to be partial." As you no doubt know, the Adams' hackle mixture involves grizzly plus red-brown; however, it's not just a matter of blending gray and brown. As an experiment I once tried dyeing the white bands in grizzly hackle to a medium brown so that I could get the dark gray and brown into the same feather. The effect I got might best be described as "muddy"; it was quite unlike the appearance of an Adams' hackle. I believe that the minute points of white versus black contrast in the banded grizzly feather blend with the red game in a way that is somehow nearly magical, and perhaps this brings us back to the blue-plus-yellow color flecks in the artist's foliage, this time fish-eye style.

It's not unlikely that the Adams is our most popular dry-fly pattern, nationwide. I routinely tell students in a fly fishing course I teach that if I were somehow limited by law to just one pattern (admittedly a bizarre thought), the Adams would be a very wise choice. Nonetheless, anglers agree that the Adams has an Achilles heel, for the pattern is often very difficult to see on the water. Somehow the Adams mix of blue-dun fur body and brown plus pepper-and-salt hackle melds smoothly into the background of surface glare and the indescribable blur of muted tints created by various swirling currents, shadows, reflections and variegated water and bottom colors. The impressionistic hackle has nothing positive to offer in terms of visibility—quite the opposite—and the hackle-tip wings get lost among the hackle fibers. I'm reminded of an unkind remark I once overheard made by a student with reference to one of his professors (possibly myself): "Man, he fades right into the blackboard!"

That comment sums up the Adams' plight pretty neatly, and the objection is a practical one, since poor fly visibility can be a limiting factor in your success. This is particularly true when the trout are "sipping and spitting." You know what I mean—a quick, delicate take with almost no surface disturbance followed instantly by expulsion of the fly. The angler has just about half a second in which to set the hook, possibly less, and if the fly isn't in sharp focus at the time, he's likely out of the game. Indeed, many strikes of this kind go unrecognized altogether. I'll bet you've watched over anglers from the vantage of a high bank and wondered why they didn't respond to obvious rises (obvious to you). I know I've been on both ends of this situation; it's just very difficult to connect while frantically searching for your fly. Of course, there are ways and means of positioning yourself in the interests of easier fly finding, such as getting up on an elevated bank; however, it makes sense to start out with a pattern that's basically easy to spot in the first place. These considerations are pertinent to a fly called the Yellow-Winged Adams.

I don't know who originally conceived of an Adams tied with yellow wings. It's not hard to figure out why he did, though, for these wings multiply the pattern's visibility many times over. I recall first seeing it advertised in the 1950s, and it's surprising that the Yellow-Winged Adams hasn't become more popular. In checking catalogs from six major tackle companies recently, I found that only one carried the pattern. (The Adams Yellow gets its name from a yellow body, not the wings.) Perhaps the Yellow-Winged Adams didn't catch on because others found, as I did, that the fly tied with bright wings sometimes attracted flighty insincere rises as compared with the regular Adams—that is, it had an unfavorable strike:hook ratio.

When the Spanish Fly's gold wings proved so effective, it was only natural to replace the Adams' hackle-tip wings with gold polypro yarn or dyed calf tail. And then there was the matter of the tail. A kosher Adams tail should be made of the same blend of hackle fibers as in the definitive hackle; however, I switched to gray-brown deer body hair because it was stiffer, balanced the heavier wings better, and liked to float on top of the surface film. Further, a deer-hair tail is not a solid color. If you look at individual fibers, the very tips are often black and then there is a pale-buff band just adacent separating the tip from the rest of the darker fiber. When the hairs are stacked to get the ends even, these bands of shading tend to overlap and stagger, resulting in a speckled appearance, even as in Mr. Atherton's paragraph quoted above.

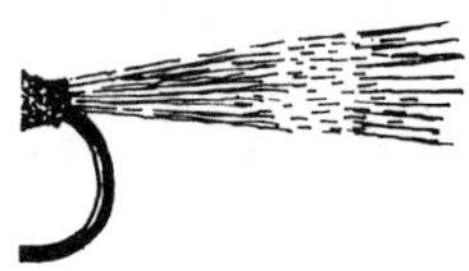

These two modifications of the Adams chassis added up to my gold-winged version. I could hardly wait to test the new pattern, but the first Golden Adams were tied in October at the end of the Spanish Fly's highly successful maiden season and it looked as though I'd have to wait out another long winter. Then I remembered that we hadn't seen my wife's family in Idaho for quite some time . . . perhaps a pre-holiday visit? The fact that they live near some outstanding trout fishing was practically incidental, and as a reward for this selfless thinking, the Golden Adams did manage to get its hackles wet—at the foot of the Devil's Bedstead, no less! Briefly, the Bedstead is a cluster of high peaks in the Pioneer Range east of Sun Valley, and from it spills a portion of the headwaters of the Big Lost River, a sizable stream later destined to sink into lava deserts near the otherworldly Craters of the Moon. It's an austere place of nearly naked mountain slopes towering above sagebrush flats cut by shallow canons.

On this bright-blue October day the frigid river nurtured globular gems of ice wherever spray from currents coated partially submerged twigs and bankside brush with layers of crystal, adding to an already striking scene. I was with my wife's Uncle Hank, a superb angler, and we split up just before noon, planning to find each other a little later to compare notes.

The first two hours couldn't have been more sterile. Dry flies were ignored, and neither nymphs nor streamers earned so much as a single strike. I headed upriver to find Hank and maybe some advice. In the process of climbing around and over a high rocky outcropping, I finally spotted him far below on the opposite bank, drying out a boot. Via sign language, it became apparent that Hank had also found the fishing slow when he held up an index finger, punctuating this bit of primitive sign language with a shrug and discouraged shake of his head.

Motioning to his creel, he reached inside, obviously searching for something small and eventually finding the tail of a trout. Slowly, slowly, out it came, 6 inches, 12, 16 (I gasped), and finally 20 inches of muscular, well-built rainbow (more like a tight end than a wide receiver, if you see what I mean). It was almost cruel.

A congratulatory wave and I was off, literally jogging up the trail at a pace that severely stressed my somewhat neglected cardiopulmonary capacity. I'd seen a Royal Wulff dangling from Hank's tippet, so a dry fly it would be. Well, it wasn't to be; ninety minutes later I had yet to see a rise, spontaneous or otherwise, after working over literally dozens of pieces of likely-looking water. Still, I wasn't whipped yet, for the vision of that gorgeous rainbow was sharply etched. The situation was dire now, as rapidly lengthening shadows seemed to streak across whole stretches of water that had been sunny just minutes before and a fresh cold wind announced the advent of evening. We were due back in Sun Valley for dinner soon, too, with the graveled Trail Creek Summit road to negotiate. What to try next?

As you've anticipated, I thought of the Golden Adams; there would be no story otherwise. I had been using proven patterns previously, but now I reached for a box with a few of the very first prototypes; in the manner of Adam and Eve, they'd never seen action. The situation was a quietly desperate one as far as I was concerned, so I remember the events that followed with considerable clarity.

Up ahead in midriver a spire of rock rose some 5 feet above the surface and had spawned a pretty triangular pool outlined by the base of the rock and merging tongues of current below. It was nice water, although no better than many spots I'd flailed with my floaters that afternoon. Even though the first cast fell short into the bouncing current at the triangle's tip, I was happy to find the gold wings almost at once as they drifted jauntily back to me. I fired the next cast some 8 feet farther into the right-sided current, but this time I lost the fly; it just wasn't anywhere to be seen. Irritated, I lifted the rod to begin another backcast, at which instant a sizable rainbow gave me a pretty strong hint as to the whereabouts of the fly by bursting sprayfully skyward. Although he measured a mere 16 inches (mere in comparison to Hank's fish, which I had by then fantasized into the 8-pound class), this broad-chested and courageous fellow used up most of my backing plus ten minutes before he was beached. It's likely perverse to strike a fish with any degree of malice in the process of killing it; however, I'm afraid I had some feelings of this kind, for it had been a very long day and both my casting arm and my pride ached. Gratifyingly, during the twenty minutes of light remaining the same Golden Adams proceeded to take three more raucous rainbow, ranging from 13 to 15 inches, all carefully released on the basis of a mixture of partially restored self-esteem and residual guilt.

I've already broken a promise made only two chapters ago about not telling fish stories, but I don't claim that this one proves anything whatever about the Golden Adams. The fly may or may not have had something to do with my last-minute redemption. Perhaps this was a very special segment of river and the water temperature was at the highest point of the day and an unseen hatch could have been responsible. There's no way to tell.

It may be that another or almost any other pattern would have fared as well, although I wasn't about to switch around after that first fish, you can be sure. I was interested to learn that Hank also ended the day with four, three like mine plus his multi-pounder, and that he caught all of them long before the Golden Adams lost its virginity.

By the time the next April rolled around, I was equipped with enough Golden Adamses in sizes 10 through 18 to open a mail-order business, they all got lost or beaten up, meaning that the pattern really worked. Over the years I've grown to like the fly so much that the regular Adams has been pretty much relegated to the laboratory as a standard against which I test other patterns.

And speaking of that sort of exercise, what happens when the gold-winged, deer-hair-tailed fly meets a real Adams in head-on competition, same day, same water, and same angler? This is one aspect of keeping a careful angling log that's fun, because, having conducted a long series of confrontations of this type, I believe I can give you an accurate report: About half the time it's a dead-even draw, with just one difference. Since it's so much easier to see, the Golden Adams requires less concentration and thus causes less eyestrain. Then, on about four occasions out of ten the gold-winged fly comes out ahead by a clear margin. I suspect that there are two and possibly three reasons for this:

1. Fewer strikes that I fail to see and hence miss. (I'm not prepared to state that the trout also see the fly better, although I suspect they do at times.)
2. Better, higher flotation because of the wings and stiffer tail. Clearly, explanations 1 and 2 are very much related.
3. At times the wings may actually make the pattern more provocative to the trout.

This leaves about 10 percent of trials when the regular Adams is the winner, usually because of a more favorable strike:hook ratio. As is my habit, each season I fish with dozens of dry-fly patterns, and because I find it entertaining I keep records on how each of them performs. During the past six seasons the Golden Adams has ranked first three times and never lower than fifth. So I'd sooner go astream without my lucky hat than my collection of golden Adamses, and I really depend on that old felt relic to get me through too.

I promised to discuss dyed wings, and will begin with the strong suggestion that you do your own. If they are purchased predyed, there's just no way to get what you want consistently, and there's almost nothing to dyeing. You don't need to get into complex chemicals; any of the popular commercial fabric dyes will give the desired result. I generally use white calf tail in the best possible grade (usually 50¢ to $1). You can pretty much tell the precise color shade you're buying from the dye package, and I often can find what I want without any mixing. In other other words, simply purchase a package of gold dye. If you enjoy fussing, however, there's plenty of opportunity; for example, mix a little brown with bright yellow.

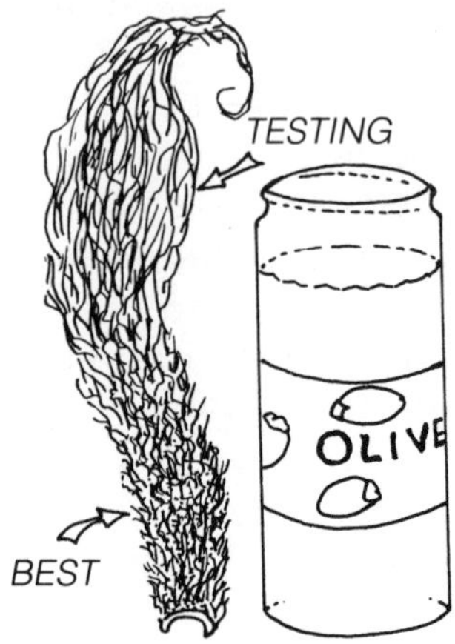

There's no need to gamble or risk ruining an entire tail, either, since you can test the color on the long wiry fibers from the tip first. The best hair for wings is in the lower half of the tail closest to the base; these fibers are finer and more pliable. I just clip off a tuft from the tip and hold it with a clothespin while trying out the strength of a dye solution and dyeing time. It's obviously important to rinse and dry the test sample to see what you've really got before dunking the whole tail. Temperature can also be a factor; the warmer the dye, the faster and more deeply it will take. I like to use a tall, wide-mouthed bottle such as those that olives come in, and when the process is too slow, I place the dye bottle in a deep saucepan of hot water. Once the strength, time, and temperature are established, the whole tail goes in, and you should soon have enough hair for several dozen flies, ready to be clipped from the skin as needed, neat and clean. The color will be fast or permanent so long as you've been careful to rinse the test samples thoroughly.

For some reason, it makes wives quite nervous when materials are dyed in the kitchen. It's hard to understand why. I practically never stain the stove or sink, at least not seriously, and prefer to work when alone.

As long as I'm making comments of a general nature that don't apply just to the Golden Adams, the question of hackle *quantity* comes to mind. Currently there is a true schism among tiers over the matter of the hackle's relative importance in the anatomy of a dry fly. Once considered to be the very essence of a floating artificial, the hackle has come under a dark cloud of suspicion. Today is the era of the sharply shorn or hackleless dry fly, sort of a cross trend to men's hair styles. Now it's the body, tail, and wing silhouette that count most, and hackle is sometimes considered as extra baggage if not a gross and objectionable part of a pattern. It has been suggested that traditional hackle doesn't look natural to trout, possibly it even offends them, and the issue has reached a point where one almost feels compelled to take some kind of position.

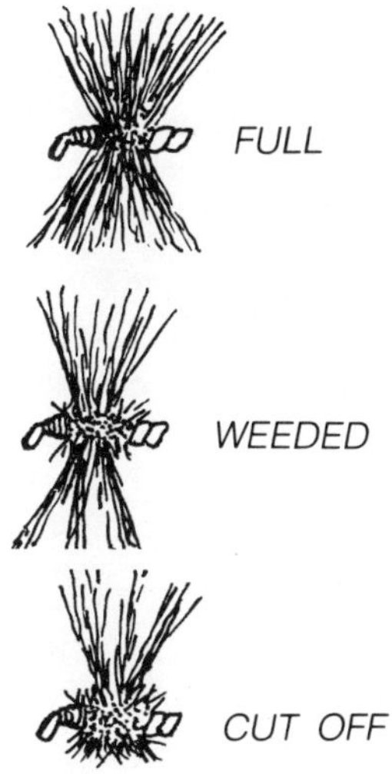

Here is what confuses me: If old-fashioned, regularly hackled dry flies are really a bad idea, why has it taken so long to find out, and why have so many anglers done so well with them over the past decades of dry-fly endeavor? Perhaps today's trout are more suspicious or selective, since they see more and more artificials floating overhead each year, but it's hard for me to buy this. A lot of big old trout that should know better still get caught on normally hackled flies, fish like the 17-inch rainbow my son Charlie took in downtown Aspen under the Jerome Hotel bridge on a hugely hackled Royal Wulff.

No doubt you can photograph the undersides of floating flies, both natural and "un," through the bottom of a glass baking dish or aquarium or from the bottom of your bathtub if you're sufficiently agile, but so what? If beauty is truly in the eye of the beholder, what right have we to judge for the trout? Would they admire magazine centerfolds in the same way as fisher*men*? I'd think not, although we might see some revolutionary mermaid lures if they did.

Anyhow, new concepts are always welcome in any field, but I wonder if we aren't overdoing it in the case of the denuded dry fly. Admitting that scrupulously perfect matching patterns are intellectually appealing, shouldn't we be sure that they really are superior lest we lapse into sophistry? After all, trout don't read the same books and articles; they may not know how to react in a properly modern manner. With so many of today's experts classifying insects as to genus and species, I suspect we perhaps should guard against being classified ourselves as *Piscator pedanticus*.

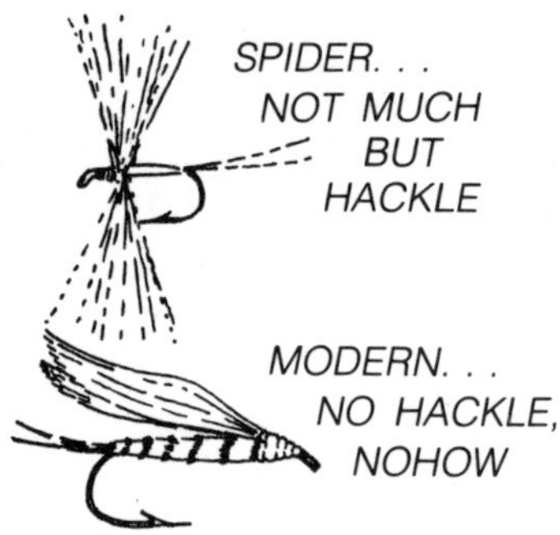

Personally, I see to it that my flies are decently dressed with hackle—not a thick wadded muff of feathers, but a reasonable radial ring of quality fibers. Then, if the need arises, I can trim the bottom flat or generally thin the whole thing as seems appropriate. It's certainly easier than trying to add hackle to a finished fly.

The idea of putting an Adams hackle onto a Spanish Fly's frame may have occurred to you, and it's a good one. I prefer to carry both patterns with me, though, because they have different "looks" on the water. The Spanish Fly is distinctly yellower, whereas the Golden Adams has a gray cast because of its dun body and complement of grizzly hackle. Frankly, I can't bring myself to break up the Adams' proven combination of blue-gray fur dubbed body and grizzly/red-game hackle mix. For one thing, it retrieved an apparently hopeless situation for me at the very foot of the Devil's Bedstead!

CHAPTER 4

The Laramie

As much as I depend on the Spanish Fly and the Golden Adams, the Laramie is my real favorite. In order to explain the whys and wherefores of this pattern, I need to go back to the Adams briefly, hoping I've not run the topic all the way into the ground by now. I grew up in Denver and, like most fly fishing families, we had a copy of Ray Bergman's classic *Trout.* It was the one book my parents could get me to read; in fact I'd go through it several times a year, often in preference to homework assignments. Having essentially committed *Trout* to memory, I knew all about the Adams, although I'd never fished with one, for I had been brought up on wet flies, so to speak. Nearly everyone has read this book—it comes close to being the fly fisher's Old Testament.

I wonder if you recall just where in the text the Adams saved the day for Mr. Bergman. It was on the Encampment River in the mica mine country of southern Wyoming, and as it happened, when I was about seventeen, my father and I were invited to fish the exclusive waters of the rustic Wyco Club on the lower Encampment, practically the same stretch described in *Trout.* I consequently stocked up on Adamses beforehand, for this was to be my initiation to dry-fly fishing. I also remember purchasing my first tapered leader for the occasion (I'd used snelled flies and looped leaders up to that point).

Although this was over thirty years ago, I still rate that trip as *the* most exciting of what has by now become a fairly lengthy angling career. In line with my agreement about not telling fish stories, I won't go into detail, and some of the trout were small anyway. However, on the Encampment the little guys still ran about a foot, whereas the run-of-the-mill fish averaged 2 pounds and the big boys were the kind you don't talk about since no one will believe you on any account.

Near its junction with the North Platte, the Encampment gets pinched into a canyon between two immense piles of rock, and one morning while preoccupied in combat with a rainbow, I was startled to see a full-grown cougar crossing a riffle not fifty yards below me, taking great care not to wet its feet more than absolutely necessary. It was an occasion I'll not forget, and returning home, I read over the Adams segment again for what may have been the forty-third time. I paid particular attention to the part about mixed hackle colors and noted that Mr. Bergman also described several other patterns which employed this tactic.

The Blue Fox interested me most. Like the Adams it had a blue dun fur dubbed body and grizzly hackle-tip wings, although this time the hackle was a mixture of grizzly and blue dun feathers, as was the tail. It was a disappointment to find that no one in any of Denver's various tackle shops had ever heard of the Blue Fox, so I decided I would have to tie my own, and I had the equipment at hand. A hitherto largely neglected Thompson Model A vise, hackle pliers, and a set of essential supplies had

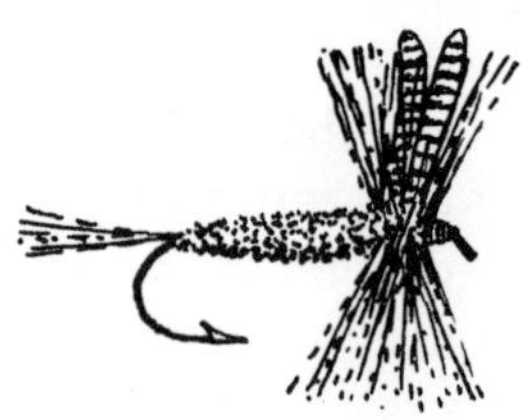
BLUE FOX

been a Christmas gift several years before, but it soon became apparant that I lacked certain basic tying skills. My Blue Foxes invariably displayed "random" wing orientation (every which way), and the lumpy bodies looked as if the flies had eaten something they couldn't quite digest.

Adams I could buy, but I was determined to also explore the possible virtues of the Blue Fox, so I dealt with my dilemma in the most practical way possible by simply leaving out the wings altogether and replacing the fur dubbed body with more easily managed silver tinsel. The result was a type of variant, I suppose, pretty remedial insofar as construction characteristics, but a surprisingly effective fly nonetheless.

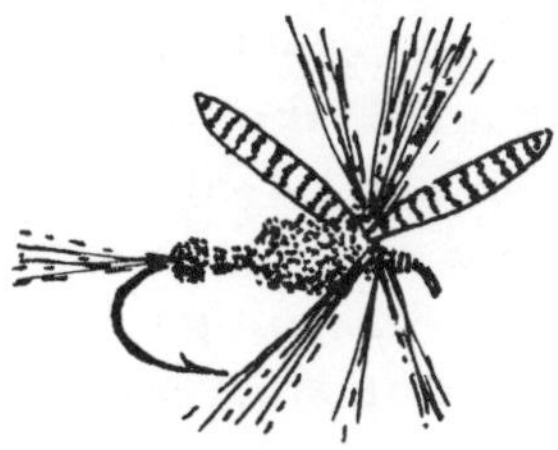

This was particularly true on the Big Laramie River, an excellent brown trout stream that heads in the high Rawah Peaks, a spur off the Continental Divide northwest of Estes Park, Colorado. The Big Laramie flows due north to (appropriately) Laramie, Wyoming, and as it descends from timbered slopes to wander through willow-thicketed hay meadows, it is the pastoral picture of a Western fly fishing stream. We visited the Big Laramie a great deal in those days, and my primitive artificial with its grizzly and blue dun hackle seemed to fit the river as a key fits its lock, opening the door to some very fine fishing. The biggest challenge wasn't so much a matter of bamboozling the browns as learning to walk through the flooded meadows along the river without breathing. It was no mean trick, either, but the alternative was quite literally suffocation by mosquitoes, a grisly way to end a fishing trip to be sure.

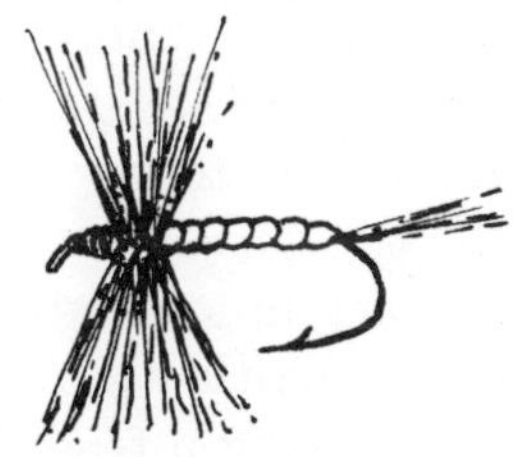

A fisherman and his fly really make up a team, and in this instance I'm afraid the stripped-down Blue Fox was the more capable member, for I was quite inexperienced with dry flies. However, the two of us still managed to catch all sorts of trout over a long series of seasons until the Spanish Fly came along.

Then the Laramie (as I had come to call the fly) sprouted golden wings along with a lot of other patterns; they were spreading throughout my collection like a disease at the time. I added a deer-hair tail, too, thereby enhancing flotation, and having copied the Spanish Fly to this extent I decided to go whole hog by putting a muff of peacock herl behind the wing base.

Now the once-trim Laramie had become a sartorial disaster. The gold wings clashed with the hackle and silver tinsel body as if it were a sort of clown fly, tied as a joke. Have you ever been put down really hard by a remark, not just an average zinger but the kind of verbal slam that sticks in the memory? The simpler the comment, the more devastating, and thanks to the Laramie I absorbed a real beauty on the banks of Silver Creek, Idaho's exotic chalk stream.

The morning's hatch had concluded some forty-five minutes before, and Silver Creek lay glassy-still, the earlier excitement no more than a lingering and wishful memory for the several anglers who remained. For want of a better plan, I was flinging a gold-winged Laramie about rather aimlessly when a nice rainbow quite unexpectedly quaffed it down in the manner of an after-dinner brandy. As I released the glistening thick-bodied fish, a shadow fell across my shoulder and I beheld a neatly attired gentleman whose hat badges and shoulder patches proclaimed origin from California. "What'd you get him on?" quoth he suspiciously.

I suspected he was looking for a spinning lure if not a worm, and I turned with a smile, for I was pleased; he had such a tailored look of expertise about him, what with his fitted waders, jaunty hat, and fine bamboo rod (mine was fiberglass). "It's just something I tie," I offered modestly, handing him the fly.

I began to fumble for my box with the intention of proffering a fresh copy as a gift, whereupon he proceeded to lay me out flat, and I quote: "The good fish quit rising some time ago; you must have found the village idiot."

Most often when given time to reflect, a reasonably literate person can come up with some sort of countercomment—perhaps not at that precise second, but eventually. To this day, I've not thought of a single acceptable reply. After all, he may have been right!

After this not-so-touching encounter, I thought seriously

about dropping the pattern altogether, or at least defrocking it back to its original wingless state, when an accident came along that improved both the pattern's appearance and its performance. As pointed out earlier, it's most important to rinse your materials thoroughly when testing for final color during the dyeing process. Otherwise the wings will fade with repeated dunkings, and I made this error in the process of dyeing an entire calf's tail. Unaware of this oversight, I tied up a whole batch of Laramies with wings that started life as medium gold only to wash out to a dark cream after a few minutes on the water. And the paler the wings became, the more trout the fly seemed to attract, until by that season's end, the pallid Laramie had whipped everything else in my arsenal.

Skeptical? I don't blame you. It's a pretty strange pattern. For instance, the band of dark peacock herl in the middle of an otherwise light-colored fly—what sense does that make? My basic intent was to add a body material with proven appeal on an empirical basis, just as in the Spanish Fly, but there may be another reason for the herl's positive contribution. If you float a Laramie or Spanish Fly in a glass baking dish and view it from underneath, you'll see that the herl band expands and darkens the body just about where the wings and legs would emerge from a natural's body. So it's possible that the herl cuff provides a bit of realism in terms of both shade and silhouette. Of course, I have no proof, and actually I'm pretty certain the fly would function nicely without the herl. But you know, the fuzzy green body band makes *me* like the Laramie better, and I think this is important. If a small wrinkle that's easy to add gives an angler increased confidence in his pattern, this is reason enough, it seems to me.

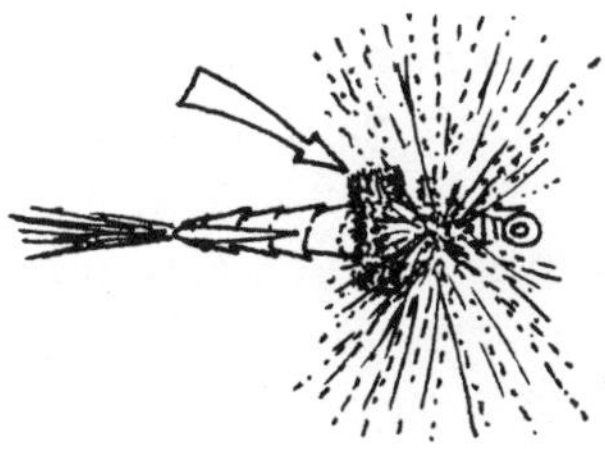

So to summarize, the Laramie was designed to make a strong statement along the lines of impressionism; the mixture of hackles, somewhat discordant wing color, dark band of herl, sparkle provided by the tinsel, and stippled hair tail all serve toward this end. The wings make the fly easy to find on the water, while the armor-plated abdomen gives it durability. Thus it's both a functional and a practical sort of artificial. And whether or not it makes any difference, the Laramie is decidedly unlike anything else the trout can have seen. If trout have a capacity for curiosity, the Laramie ought to arouse it, and it may, too, for in my experience they will usually come up for this pattern if they're willing to rise at all. In short, the Laramie seems to get their attention!

Of course, I think of the Laramie when there are grayish naturals about, although there is another and much more common situation that makes me reach for this pattern. I'll explain a little tangentially.

Over the years, I've taken a fair amount of guff about my fishing log, or "Dad's dumb diary" as it's known in the family. Perhaps it is silly for a grown man to slavishly record each day's angling activities in some detail, but this is one compulsive habit I'll never give up; I've learned too much from my so-called diary. You'd think a zealous angler with most of his mental machinery still intact would have accurate recall for important streamside events without recourse to a log. However, in my case it doesn't work that way, for I routinely get things all muddled. Dates, places, water conditions, insect activities, and which fly did what sometimes meld into an indistinct blur; or worse, memories I'm sure of prove to be erroneous when checked against the record. The significance of identical or similar observations can be lost if those observations are spread out in time unless they are somehow recorded.

For example, it was through gradual accumulation of data and repeated review of my log that I recognized something quite useful about the Laramie. Briefly, while the pattern is a steady performer under most conditions, when skies are cloudy, it can be really exceptional, even during a storm at times. This has been an across-the-board phenomenon that has nothing to do with trout species or the presence of naturals; an overcast leaden sky just brings out the best in the Laramie, and I have no idea why. In any event, during the past nine seasons the cream-winged Laramie has ranked among my top four dry-fly patterns eight times, rain plus shine.

I should explain that the Laramie's particular mixture of hackles can be tricky to handle. As you know, the term "blue dun" covers a tremendous range of shades, from a very pale feather to iron-blue dun, which is virtually navy blue. Now, if light or pale dun hackle feathers are mixed with the grizzly, the effect is that of an almost white hackle peppered with the black bands of the grizzly, and I've not done so well with this combination. Similarly, I can't recommend a mixture of iron-blue dun and grizzly, because this gives a black-appearing hackle, this time salted with the grizzly's white bands. I believe that a medium-blue dun feather gives by far the smoothest blend, yielding a subtle mottled gray look that I and the trout (I think) find most pleasing. Beyond this, I understand that in painting there are certain strong colors or combinations of colors and others which are weak or recessive such that the former dominate in a picture if given a chance. In the case of the marriage between blue dun and grizzly it's the grizzly that has the clout. For this reason I'd suggest a 60/40 or two-thirds to one-third mix in favor of the dun component—say, two blue dun feathers to one grizzly or a long dun feather plus a short grizzly hackle. It really does make a difference.

PALE DUN

SAME AMOUNT OF GRIZZLY

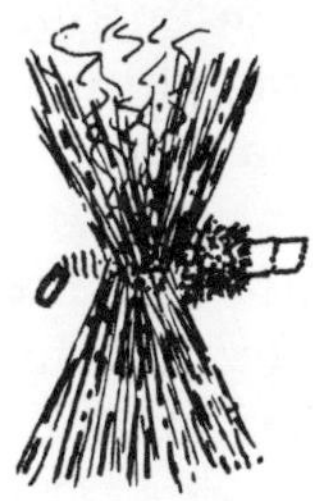

IRON BLUE

Unfortunately, the Laramie calls for two expensive kinds of hackle. High-quality grizzly isn't hard to find if you're willing to pay for it. However, natural blue dun necks are currently priced in excess of $30, and if you spend less for a dyed neck, it's all too possible to get a mediocre bunch of feathers from the standpoint of either color or dry-fly-hackle quality. On this account I got into photodyeing recently and would encourage you to give it a try. There's an initial outlay of at least $6 for a top-quality white or off-white neck, but thereafter the process is neither costly nor difficult. Since photodeying is essential in the construction of my Tadpole pattern, I'll save the details for Chapter 11.

I'd like to include a brief description of a variation of the Laramie that's designed particularly for big rough water. I call it the Roaring Fork Special because it was developed on the descriptively named Roaring Ford of the Colorado, west of Aspen.

Once it picks up sizable Hunter, Castle, and Maroon Creeks, the Fork becomes a tough, rampaging river, and in it live some equally rugged rainbow. These big silvery fish frequent heavy current rips when they are feeding, and fly drownings are commonplace in this welter of white water. In addition, casts are often lengthy, so the best-possible fly visibility becomes important. In order to improve both visibility and flotation, I replace the tinsel-and-herl body with clipped, shaped deer hair, just as in the familiar Irresistible dry-fly series, retaining the hair tail. This body construction gives a high, cork-like float, and when the fly is finally soaked, the hair can be squeezed dry as in wringing out a sponge. A quick dip in silicone dressing and you're ready to go again. I also tie the hackle a little thicker and shorter than is usual for a dry fly and beef up the tail into a veritable shaving brush.

BORN TO FLOAT

It's a buoyant bug, all right. One afternoon on the Roaring Fork several years ago, a single Special demonstrated remarkable stamina and stability by catching sixteen consecutive rainbow, all of them quite angry about the whole thing. The fish were so aggressive and difficult to pacify that I overstayed my time limit and was tardy for a social affair back in Aspen as I hurriedly forded the river to regain the trail. At times, the Fork's polished rock bottom gets to be like soaped ice, and in haste I grew careless. It was a good float and a free float, none of that splashing about trying to get your feet back under you. I was, in fact, rafting without a raft, and my head pointed downriver where by tilting my chin way back I could see the Woody Creek bridge bobbing against the blue sky as I sped along. Several boulders whizzed by, perilously close to my head, and I'd covered most of 200 yards before one knee slammed into a rock hard enough to stop me. Still clutching the rod, I staggered painfully to the bank, reflecting dazedly that the top of my hat (which had somehow stayed on) was the only dry part I had left and that my fly was somehow still floating! Sure enough, the tattered Roaring Fork Special was bobbing happily at my feet, its saucy wings erect. This would be an outstanding fish story if a lunker had happened to hook himself during my voyage, but also an untrue one, I'm afraid.

And speaking of hair wings, I've stated repeatedly that they add buoyancy to an artificial without explaining why. I think the most important factor has to do with the relationship between the *base* of the wings and the surface film. The wing roots form a wedge (essentially two cones pressed together) of low-density material that pushes the surface film downward without breaking it. It's not hard to see this if you look closely. As the wings splay away from the shaft, the surface film around them dimples down even after the bottom hackle fibers and underside of the body have submerged. Thus, while the fly is no longer sailing along up on its hackle tips, it is still floating and will continue to do so for some time. In swift, rough water I've not found this sort of half-mast presentation to discourage the trout in any way. Instead, it many times seems to increase credibility.

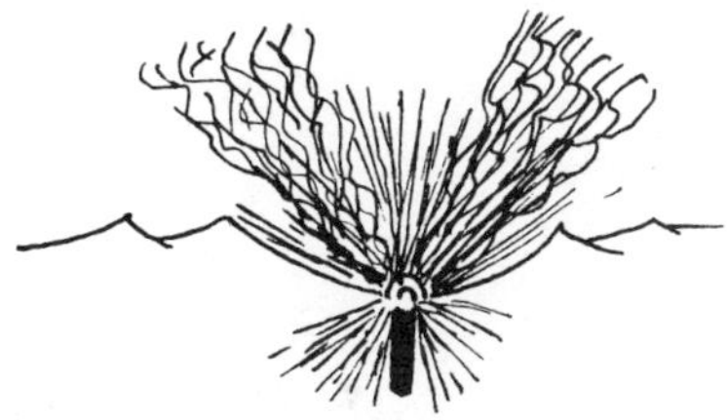

VULNERABLE

I'd like to conclude with a humble suggestion about tying hair wings. I used to have terrible trials with hooks that broke at the bend on my hair-winged artificials; more of them were ruined in this way than were worn out or lost. It seems pretty certain that this focal weakness in the metal was the result of trauma inflicted during my heavy-handed or otherwise inept efforts to cinch down the wing materials. When a hook is held in the vise in the usual position, the pulling and tugging gets transmitted to the bend just where it emerges from the vise jaws. It's the same thing as breaking a piece of wire by bending it back and forth. I find it hard to be very gentle when putting on hair wings because calf hair is such wiry, springy stuff. By the time I've covered the hair butt ends, gone through the X-wrapping to separate the wings, and put on enough thread to get them to stand up

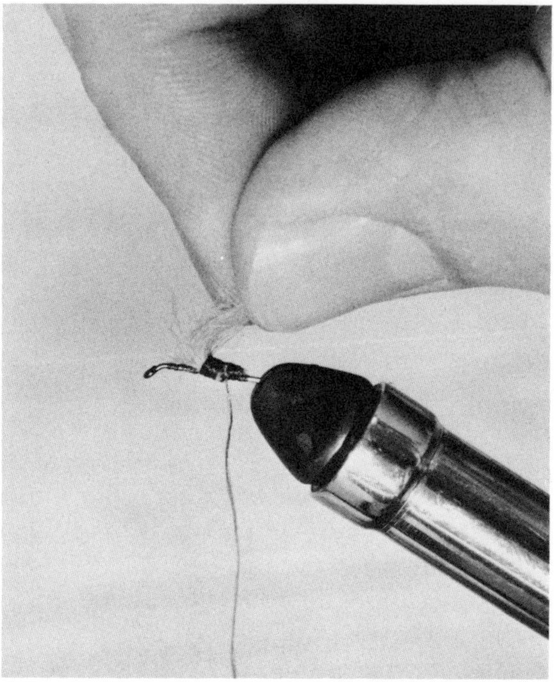

In tying the Laramie, the fingers hold the hair wings erect while the hook's shaft and bend are protected by the jaws of the vise.

properly, the hook will have absorbed sufficient damage at that specific point to become brittle. Later it's likely to snap on a strike or sometimes when just rapped against the rod shaft during a casting mishap. I've learned to virtually eliminate this catastrophe by tying the wings first thing, before the tail and body go on. The trick is to protect the hook's bend by hiding it between the jaws of the vise while the wings are being tied. This transfers the trauma to the straight shaft, and this part of a hook is very tough; I can't recall seeing a hook that broke in its shaft. After the wings are finished, I just reposition the hook normally and tie in the tail, body, and hackle as always.

CHAPTER 5

The Maniac Nymph

The ultimate achievement in fly tying would be the creation of the perfect nymph. It's the tier's Holy Grail, a fly that by definition would catch trout in water of any kind under all conceivable conditions.

Why a nymph? Why not the perfect dry fly? There are several reasons. For one thing, several strong claimants to the dry-fly title are already on the scene, and if these aren't perfect patterns, they at least qualify for "super" status. If you went to the trouble to get a large number of anglers throughout the nation to list their five favorite dry flies and mail in the ballots, I'd bet that the Adams and Royal Coachman/Wulff would stand on the very pinnacle of patterndom. But if one were to try to do the same thing with nymphs, I suspect the result would be an uninterpretable mess.

MONTANA NYMPH

ROSBOROUGH NYMPH

WOOLY WORM

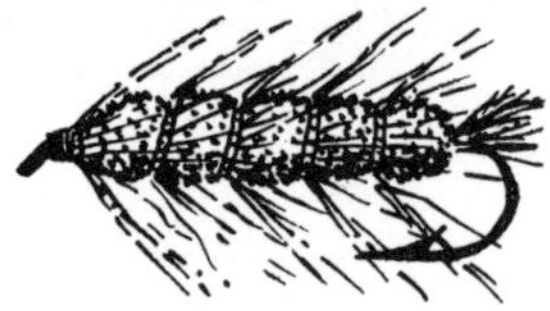

To begin with, I'd predict mass confusion as to whether standout patterns such as the Hare's Ear and Grey Hackle Peacock represent wet flies in the sense of drowned adults or nymphs and whether or not it's worth worrying about the distinction. On top of this debate there would be well-defined pockets of regional popularity, as in the case of the Montana Nymph. This black-bodied fly with its yellow palmer-hackled thorax should score heavily throughout the Rocky Mountain states; however, I'm not so sure its popularity would hold up elsewhere. Meanwhile, Polly Rosborough's Fuzzy Nymphs would also get a lot of well-deserved attention, although his is a whole family of flies, structurally related, no doubt, but still distinctive patterns. As a result I'd expect votes for individual members to get diluted, and the same would be true for the ubiquitous Wooly Worm, some ballots in favor of black and others for brown, yellow, orange, olive, and so on. After the results were all in and tallied, you could go on to construct a ranking list, although I think you'd still get a blur of names, none of them standing out very clearly from the others.

There's a much better reason to search for the perfect nymph. Whether wet or dry, the ideal fly would have to suggest the widest possible variety of natural food forms to the trout, and when you think about it, this paragon of patterns should actually be preferable to the real thing(s). In this regard the nymph's row is by far the tougher to hoe, relative to a dry fly, because of the enormous diversity of underwater food types as compared with floating forms. Dry flies need to simulate hatching naturals, consenting adults caught in uncompromising positions on the surface, ovipositing females, and, finally, spent and senile members of the population at the end of their lifespan. Although these are very different activities, in each case a creature with the body, tail, legs, antennae, and wings of a "fly" is involved. No doubt the things that fall in from the bank and get eaten on the surface are more various, but still, any of the insects dry flies imitate may also appear beneath the surface for the nymph to suggest in addition to all of the sorts of things trout covet that by nature live underwater.

Just think how many there are. Ten major classes of common aquatic larvae are listed in Ernest Schwiebert's comprehensive book *Nymphs,* not specific insects but whole groups of related creatures such as the mayflies, caddis flies, and stone flies. And huge differences in size, shape, and color exist between members of each group, so the possibilities for variation are nearly endless. Add to these the snails, freshwater shrimp, leeches, earthworms, and tiny minnows with which our hypothetical nymph must compete and I think the point is made.

Another thing, admitting that floating insects aren't always easy to see, the dry-fly angler has at least a decent chance to identify what's on, over, or along the water. Meanwhile the purveyor of sunken artificials has the nearly impossible task of spotting a veritable menagerie of morsels crawling among the nooks and crannies along the bottom or maundering about the murky depths of pools and slack water. He can only guess at what's down there, and thus a pattern with general appeal is all the more valuable.

It isn't that we don't already have effective nymph/wet-fly patterns; there are plenty of them, and some go all the way back to the very beginnings of recorded flyfishing history. The structurally unsophisticated patterns so entertainingly described by Sylvester Nemes in his *The Soft-Hackled Fly* are good examples and are probably every bit as effective today as they were a hundred years ago. It follows that it's quite easy to create "new" flies that will almost certainly trick the trout by just modifying these good old patterns slightly. You could go on to give them names, write a book about them, and generally tickle your ego; however, that would be a deceitful waste of the reader's time. Thus, I feel that in order to qualify as new, an artificial should be significantly different from any established pattern and not just a minor variation on some successful theme.

The evolution of the Maniac Nymph required about two decades, but to alleviate reader boredom I've condensed the whole process into a couple of pages. My first decision involved the question of relative shade, i.e., light versus dark or in between. Since you can't have it all ways, I opted for the middle-of-the-road approach in the interests of generality and began to work medium browns and grays.

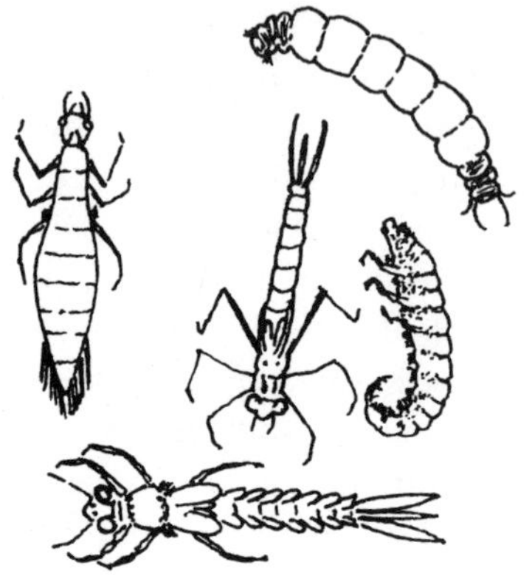

Next was the matter of shape, again a three-way decision, short and stubby versus elongate or something intermediate, and this time I went for the extreme. It's a straightforward biological fact that stone fly, dragonfly, damselfly, alder fly, crane fly, fishfly, and dobsonfly nymphs have extended wheelbases, not to mention caddis pupae, free-swimming larvae, and nymphs of some of the most important Mayfly order. Thus, I decided to build the fly on the skeleton of a 3X long hook. I used either muskrat or fox fur to get a medium-blue dun or reddish-brown dubbed body plus a skimpy beard hackle and tail of soft feathers. This all-purpose fly had a perfectly awful acronym: I called it the Camas (*Ca* for "caddis," *ma* for "mayfly,"and *s* for "stone fly"). And the Camas performed about as its name suggested, none too well. I learned one thing, though: the gray version seemed better than the brown, although in either case, my plain-Jane Camas lacked class.

THE CA-MA-S

MARTINEZ BLACK

Of course, I fished with quite a number of standard nymphs while all this was going on, and I had three special favorites. One of them was the Martinez or Martinez Black, popular in the Yellowstone area. This somber nymph has an all-black abdomen and thorax, which I tied with black-dyed angora, as per the recipe. This was still a long time ago when anglers first began to write about seal hair for fly bodies. The stuff was supposed to give an imitation of lifelike quality. Seal hair was described as both "lustrous" and "luminous" and was said to impart a vibrant sheen to an artificial. While I certainly had nothing against seals, I never really believed all this. It sounded too much like a shampoo ad. Still, I knew that the ever-dependable Hare's Ear featured a shiny shaggy body of coarse hair fibers, so I decided to at least try dyed seal on the Martinez Black. This modification was such an immediate success that I gave away all of my original Martinez nymphs tied with angora, several dozen of them, and have never regretted the decision.

MARTINEZ—SEAL

Now I began to have some really good days with my Camas tied with gray seal's hair, except that the stuff tended to shed when chewed upon. Since the nymphs I liked best all sported a tinsel rib (Zug Bug, Atherton, and Martinez; the Hare's Ear, too, if you want to count it as a nymph), and since ribbing would help hold the hair on longer, I dressed up the fly with a thin rib of silver tinsel. Actually, Mr. Atherton's book had a lot to do with the seal's hair and the tinsel, too, for he used both in the interests of impressionism. Seal hair was a particular favorite. To quote, "It has a wonderful sheen in the water as well as a very soft silhouette on the fly." To paraphrase, he felt that tight, hard bodies such as those you get with packed, soft fur dubbing, yarn, or silk are too thin and opaque. Atherton preferred a fuller, slightly translucent body of softer outline, and my experiences with the hair-bodied Martinez convinced me that he was right. As to the tinsel, he said, "It is a great help in keeping the dubbing from being worn off; gives a slightly segmented silhouette; and by its tiny sparkles it suggests the shimmer of an insect."

We aren't through with Mr. Atherton yet. Although it was a later addition to my pattern, the bright-blue wing case is the epitome of an impressionistic touch. Certainly no one ever saw a natural nymph with a shiny blue wing case, but that isn't the point. This color touch is only meant to suggest the reflection of light from the surface of the real insect. The Martinez Black displays a flash of bright green on the top of its dark thorax, presumably with this purpose in mind, while Atherton's medium nymph (properly the Atherton Medium) uses bright blue against a background of pinkish beige in the same way. So the glossy seal hair, tinsel, and colorful wing case all go into a brew of impressionism, and each plays a role in simulating the natural's shiny reflective surfaces or some component thereof. This might also apply to mimicry of tiny reflective air bubbles adherent to an insect's body, for instance, a female caddis diving beneath the surface to lay her eggs.

The colored wing cases are made of natural or dyed feathers tied on top of the thorax, and here I encountered a problem, for feather cases tend to be quite fragile. On this account I set out to find a synthetic substitute and came up with what I thought was a really brilliant innovation. I needed a material that would be strong yet pliable and that I could trim to shape. Why not the thin lead sheeting that covers the cork and neck of wine bottles? I found I could tie a beautiful fly with the metal case, and it also added just a little weight to help the fly sink. And, if the need arose, I could always put a rival tier down by coolly remarking that my nymph had a touch of Mouton Rothschild 1962 in it! How's that for snobbery? (This would be a red wing case, although it's unlikely the other party would know.) Unfortunately, it turned out that the lead soon lost its color and became a dull (leaden) gray, so I searched on. One day I noticed a neighbor patching a tent with the special tape that's designed for closing tears in nylon. It was perfect in every way, even the color, and I'm still using the small piece he gave me. I suppose in a way it's a contradiction to mix a synthetic material with natural fur and feathers; however, I didn't ponder the issue, for the tinsel rib was hardly natural, and besides, the blue-thatched fly caught trout in such frenzied fashion that I began to call it the Maniac.

We've come a long way without saying anything about legs and tails. I decided early on that mottled grayish hackle fibers would be the logical choice to match the body. John Atherton tells us that legs, antennae, and tails of real nymphs are seldom homogeneous. Rather they tend to be speckled or variegated, and the beautiful color plates in Schwiebert's *Nymphs* bear this out. So from the beginning I used finely banded guinea fowl, which, although basically black-and-white, has a gray appearance overall. Finally, as Sylvester Nemes points out in his book about soft-hackled flies, it's vital to choose pliable fibers that will react in the current by waving hither and yon in realistic fashion.

(Some guinea fowl feathers are heavily banded and stiff, and these are not so desirable.)

Even on those rare occasions when I have a good pattern recipe going, I can't resist adding wrinkles here and there, and this is how the Maniac came to have a yellow underside. With the intention of lightening the bottom of the abdomen, I tied on a strip of yellow yarn beneath the ribbing, although in retrospect, the yarn's contribution probably has little to do with color. As I've intimated, seal hair is about the shaggiest of all possible dubbing materials, and as it fluffs out from the body the veil of fibers largely hides the yarn strip. At the same time the wetted yarn also fuzzes along its edges until the two blend into one another so that the yellow is seen rather vaguely through a fringe of blue-gray hair, a misty mix of color. More significantly, the strip of yarn changes the body's configuration in a subtle and intriguing way. I find that the yarn keeps the hair on the underside of the hook from flaring, whereas the fibers along the sides of the body are free to fray and fluff. This gives the body an oval, flattened shape, which can be accentuated further by dabbing a little head cement into the hair on top of the hook to hold those fibers tight while picking out a frayed edge of hair on either side with a stylus. This fillip adds a nice touch of realism, since so many natural nymphs are somewhat flattened top to bottom, and I like to think that the lateral translucent fringe simulates abdominal gills, as in the mayfly nymphs. The reason I tend to believe that there may be something to all this is that the more battle-worn a Maniac gets, the more battered and frayed, the more effective it becomes. It has always seemed that the old veterans in my collection become wise in the ways of the trout, and I hate to lose a well-seasoned Maniac, for then I have to break in a raw rookie. I've even dared to run a batch through the rinse cycle in my wife's washing machine, a hazardous venture on several counts.

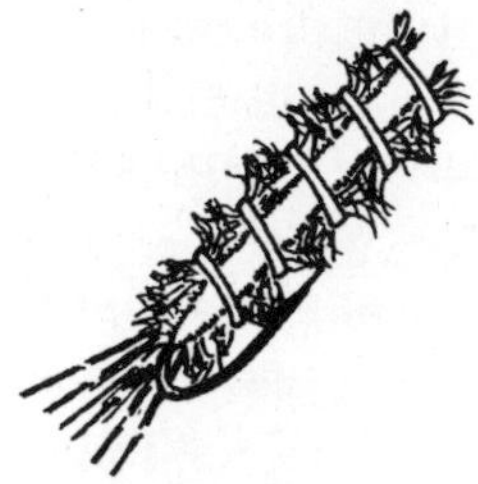

REAR VIEW—ON-END

What do the trout think the Maniac is, anyway? I've pondered this question for a long time, and not just as an intellectual exercise. Obviously, if we could understand the answer, as it applies to this or any other pattern, it should enable us to create flies that would seldom fail; indeed, they might take the fun right out of fishing. As I've explained, however, the Maniac wasn't meant to copy any specific food form, and I suppose if the trout took it well, but *only* as an imitation of a *certain* insect, the thing would be a failure. So I'm rather pleased to report that I haven't the slightest idea how they perceive the Maniac. More often than not there is little or no correspondence between what the trout are eating and this fuzzy fly except perhaps for size. Take the stone fly nymphs that wash about during the runoff or after heavy rains. For the most part, these are dark-brassy-colored creatures, and while the trout are usually quite decent about accepting correspondingly dark replicas, the paler, gray Maniac frequently outcatches better imitations matched carefully for color. In this instance, I would try to match for size, however—i.e., an inch-long natural nymph with a Maniac tied on a 3X long No. 10 hook. And then there are dry-fly situations when, although the trout are surface-feeding, they will repeatedly nail a Maniac, drifted shallowly overhead. I've wondered why, for it's rare to find anything in their stomachs that remotely resembles the artificial. In such situations there's usually no need to match hook size exactly with whatever they are taking off the top. So, in truth, there isn't much hatch-matching science behind the Maniac. It just works!

It's a funny thing, but after *Flyfishing the Rockies* was published, my local angling fame expanded far out of proportion with deserved reality. Just all sorts of folks wanted to go astream with me (except my kids—they weren't the least bit impressed; they knew). I've enjoyed the people I've met this way. Among the most memorable were a concert pianist and his lovely wife, a ballerina, with whom I fished in Aspen. Artists in their own spheres, they had never stood upon the piscatorial stage, so this trip was a first for both of them. It was well after noon when we got under way, and I chose the Roaring Fork River above town, where the stream is smaller, to save time and prevent serious mishaps in the event of wading accidents.

Intent in conversation, I picked an erroneous nontrail down to the river from the Independence Pass highway (the turnoffs do look pretty much the same), and when we finally reached the bank, scratched and winded, after a precipitous descent through the brush, I felt that the "white hunter" was one down already. In the process of getting all of us rigged up, I suddenly went two down when the lady casually mentioned that she had dinner guests arriving in four hours, that she had promised fresh trout for the entree, and that there was an excellent white wine chilling for the occasion.

Still, it was a fine afternoon and I was relieved to learn that there were only two guests, so I continued to exude quiet confidence, as befits a true expert. That all changed, though, for I quickly discovered that the trout were by no means anxious to oblige; after an hour, I'd taken but two bassinet-caliber brookies that even a starving person would have been ashamed to kill. My friends weren't in the least concerned, however. I really think they believed I could cause the fish to rise by simply pointing an index finger, biblical style. Wouldn't that be a neat trick?

In any event, ninety minutes later, we still had only two keepers plus a thoroughly wet pianist who had undergone an angler's special baptism while the ballerina had understandably turned to meditation and was seated on a rock in midriver. Things looked grim, and horrible thoughts of frozen fish from the market crossed my mind. Was the lady saying a silent prayer? My reputation was very clearly at stake, and with an apology, I hurried on alone to heavier water upriver where streamers had attracted some fair rainbows in the past. Nothing doing!

Now less than half an hour remained. I'd given up on dry flies by then, despite ideal water conditions, and so the Maniac finally got its chance. Would you believe nine trout in twenty-three minutes? It *was* like walking on water! Most were nice fish, too, so I fulfilled my commitment without exceeding the limit while regaining hero status in the bargain. They thought I'd just been holding back as a sort of joke, like coming from behind with a grand-slam home run in the bottom of the ninth. I'd hate to perform for them on the piano, though. I can't imagine salvaging a concert with a few brilliant bars right at the end. Of course, they wouldn't believe that it was the Maniac that deserved the credit, but I did.

The Maniac isn't hard to put together, although it's a little busier than most nymphs because of its belly strip and wing case. As noted, I use a 3X long hook such as Mustad's 9672 in sizes 4 through 16; there's too much detail to fit comfortably on a size 18 hook. If you're going to weight the hook, lead wire is wrapped in tight coils along the shank, overwrapped with wide spirals of tying thread and lacquered; it will never slip. The whole question of whether it's a good idea to weight wet flies is an extremely important one in my opinion, so much so that several paragraphs are devoted to this topic in the upcoming chapter. For now let's leave that decision hanging in midair and go through the other construction steps in sequence:

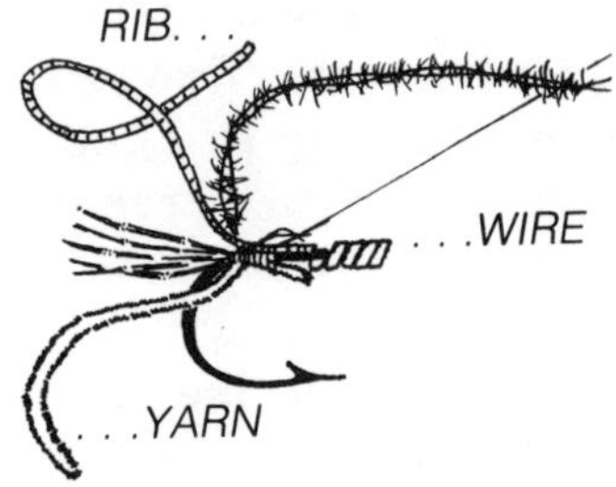

1. There are four materials to be tied in at the rear of the shank, beginning with the guinea fowl tail (see illustration for length and thickness), which goes on top of the hook. Next the yarn strand goes on the underside of the shaft, with its free end protruding to the rear. Finally, the tinsel (fine-caliber oval silver with thread core) and end of the dubbing strand are tied in on the far side of the shaft with the tinsel in front. If you've weighted with lead wire, trim the ends of all materials flush with the end of the lead wrap, cover with thread (to fill the gap between the shaft and wire), and advance the tying thread to mid-shaft.
2. Wrap on a single even layer of hair dubbing to a point just past mid-shaft and leave the dubbing thread hanging by the weight of a hackle pliers. There are lots of dubbing methods that work. However, I'm a tacky tier, if you will, and for seal hair I lay out a half-inch-wide strip in an elongate mat on my pants leg. Then a length of gray cotton thread is coated lightly with wax. I use Thompson's #3, and the thread should be just over twice as long as the strip of hair fibers. Then half of the thread is pressed down gently on top of the hair lengthwise and picked up. Most of the hair will adhere, so it requires no sleight of hand to double the thread over, sandwiching the

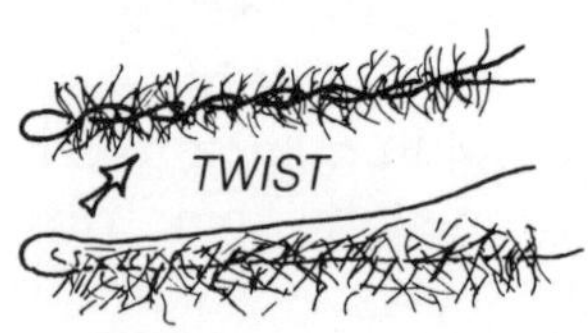

fibers and twisting the thread ends to form a double strand. This is very durable shaggy dubbing that "afros" nicely without shedding.

3. Keeping the yarn centered under the body, stretch it forward under tension, tie off, and trim just in front of the hanging dubbing.
4. Tie in the rear end of the wing case, top side facing rear. This part is fun, because you can play with the width and shape of the wing case. The most delicate and simplest case is just a thin rectangular strip. If a more obvious contribution is desired, cut a wider strip with tiny tabs at the ends where they are to be tied down; otherwise the ends will gape and bulge. When I'm trying to suggest stone fly or certain mayfly nymphs with broad, flaring wing cases, I trim to a triangular shape with a rear tab.

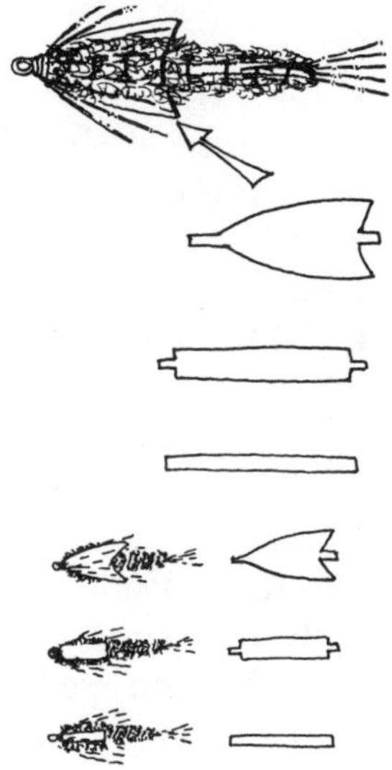

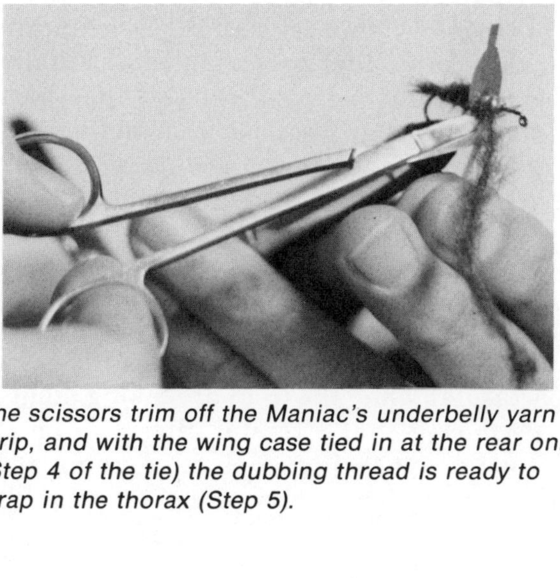

The scissors trim off the Maniac's underbelly yarn strip, and with the wing case tied in at the rear only (Step 4 of the tie) the dubbing thread is ready to wrap in the thorax (Step 5).

This actually alters the nymph's silhouette as seen from above or below, since the case will be wider posteriorly than the body. At times I think this fat, shaped case really has enhanced the Maniac's effectiveness.

5. Leaving the front end of the case free, advance the working thread to the eye area and finish the thorax with dubbing wraps, taking the first turn just behind the case (thus covering the tie-offs of the dubbing, yarn, and wing case) and the others in front. I like a plumper thorax than abdomen, so I crowd the dubbing wraps in front and tie off, taking care not to get too close to the eye.

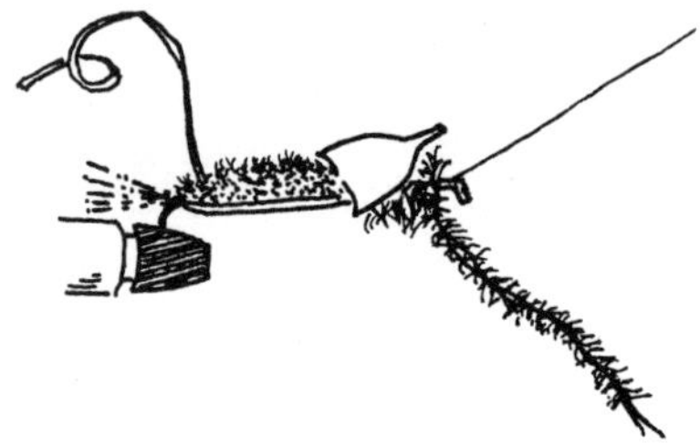

6. Now rib the entire body, abdomen and thorax, while dodging the still loose-ended wing case as you pass by; tie off the ribbing.
7. Pull the wing case forward firmly and tie off securely, trimming any excess, and tie in the tip of a soft, finely marked guinea fowl feather.
8. Wrap on the guinea fowl so that the fibers stream backward parallel to the shaft. The hackle should be sparse. Tie off and whip-finish. (Note that a lot goes on just behind the eye—the dubbing, ribbing and wing case are tied off, and the hackle is tied in and off—so leave enough room behind the eye for a tapered whip finish that covers all this.) Finally, I pick out hair fibers along both lateral sides of the abdomen with a stylus and do the same for the bottom and sides of the thorax to plump it up. The top of the abdomen gets a thin coat of lacquer together with the head, and that's it.

As I see it, the Maniac's strength lies in two related qualities. First, its productivity is not necessarily dependent upon specific mimicry. It isn't necessary for the trout to be gorging themselves on a particular natural that the Maniac resembles. For instance, it can be spectacular when the trout are taking adult flies off the surface or, as during that stressful afternoon on the Roaring Fork, when they aren't taking much of anything at all. Second, the Maniac has excellent season-long acceptability and the ability to produce in any kind of water, whether high or low, roilly or clear. And the brainy brown takes it just as well as the oft-reckless and rapacious rainbow.

Be that as it may, in the interests of honesty, I do hereby publicly admit that the Maniac Nymph is *not* the Grail Fly. The crusade to find the perfect nymph must go on, and it always will, for there is no such thing. The Maniac is really an impressionistic impresario made up other tyers' individually successful ideas. Included are the shaggy silhouette (Hare's Ear), seal hair (Atherton and others), tinsel (lots of good patterns), bright wing case (Martinez and Atherton), and soft hackle (most recently Mr. Nemes). So it's a hodgepodge, but the Maniac is also my best shot, and I guess I won't apologize.

CHAPTER 6

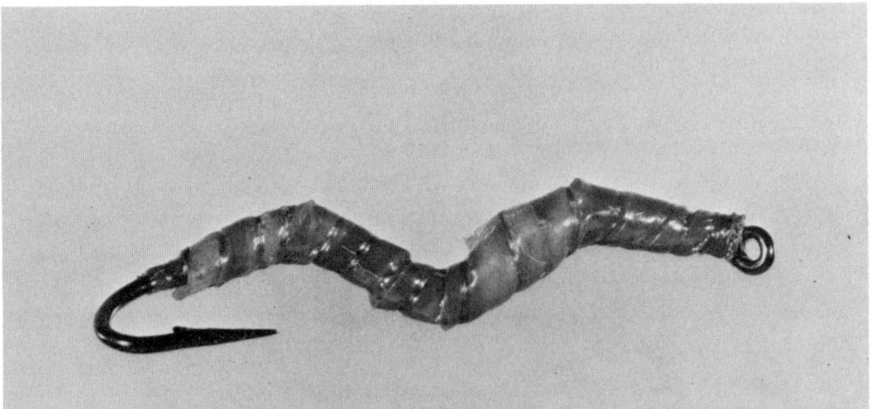

The Trashliner Flies

This chapter was more fun to put together than any of the others because it describes a novel construction material that's applied in an unusual and yet simple way. The stuff couldn't be easier to come by, is absolutely free, and has considerable potential for experimentation, so I hope you'll find the idea provocative.

It all started with an incident I recounted in *Flyfishing the Rockies.* I wanted somehow to encourage readers to get involved in fly tying, and my pitch went something like this:

I had encountered an urgent need for an artificial fly that suggested a peculiar caterpillar that was currently prevalent and therefore of intense interest to the trout. Nothing in my tackle box came close, but since I *was* a tyer, it became possible to devise an effective copy right there on the spot. Not a very imaginative approach, perhaps, but the story was even true! In brief, my myriads of green caterpillars were swarming over the willows along stream banks and occasionally falling in, much to the delight of the trout. The trouble was that I had nothing in my collection of about a thousand flies I carry around in a suitcase (only on fishing excursions) that came close to resembling the pale, slick-skinned worms. While pondering this challenge, the green plastic liner of a trashcan attracted my attention, since it duplicated the caterpillars' color almost perfectly. Padding a long-shanked hook with white yarn to achieve the correct degree of plumpness, I overwrapped it with a narrow strip of the green plastic to form a soft smooth skin. Unlike the vast majority of my creations, this one really came through.

My wonder worm was far from an everyday success, since caterpillars are both regional and seasonal. I understand that some species show up in numbers only in cycles separated by quite a few years, and it was half a dozen seasons later when I again thought of the fly. This time a murky olive caterpillar that's fond of defoliating whole aspen groves in the southern portion of the Rockies was the target. Drab green is a popular trashcan-liner color, so I had no trouble in coming up with a nearly exact match, and once again, the artificial proved to be an artful deceiver. Since then I've found that this technique has innumerable applications. The basic process involves covering some kind of body base with skin of thin transparent or translucent plastic in one color or another. There are at least five potentially advantageous characteristics of this construction.

First of all you have an interesting opportunity to blend the color of the body base with that of the plastic skin. Since either material is available in almost any hue and shade, the possible combinations are almost limitless. In the case of my first caterpillar I needed a pale milky green, and this is precisely what I got by combining a white base with the slightly cloudy light-green liner. Not long ago I met a smaller worm, a leaf roller that was falling off the bankside brush. However, this time, while the basic color was green, there was a distinct yellowish cast. Solution? There are two things to try, and you can pick the result that looks and works best: Either cover a light-yellow body with green plastic, or reverse the situation with a light-green body base and clear-yellow skin. It's very unlikely that the two schemes will give quite the same appearance, and the trout very well may show a preference. As far as color goes, there's even more latitude built into the system because the color *depth* contributed by the plastic skin increases with layering or overlapping. This is a nifty means of stimulating segmentation; you cover the body with a single layer of liner while slightly overlapping the edges as shown.

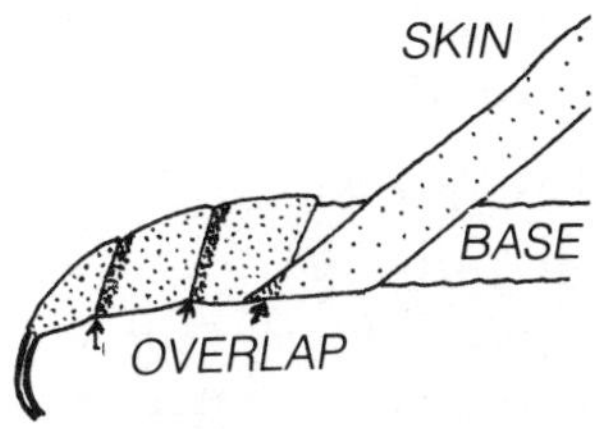

There's a second aspect that appeals to me greatly, and this is the semitranslucency of the body surface provided by the liner skin. Since the skin is either clear or partially so, it allows light to pass through the very outside of the body, yielding an impression of depth, a sort of three-dimensional effect that theoretically ought to confer some degree of realism. I often use polypro yarn for the base, and in lighter colors this material is also slightly translucent, thus enhancing the illusion of depth and smoothing out the color blend as well.

Then there's the matter of consistency. Now, I'm not going to try to tell you that trout mouth their food before they swallow it, but are we sure that they don't at times? I understand that expert bait fishermen resist the urge to rear back on their rods when they feel that first tap. Instead, they've learned to wait for the fish to get into the main course before setting the hook (and I admire them for this—my self-control isn't up to it). I don't really know whether the relative softness or hardness of an artificial's body has any effect on its acceptability, although I remember that both of my caterpillars had a superfical consistency very much like that of the real worm.

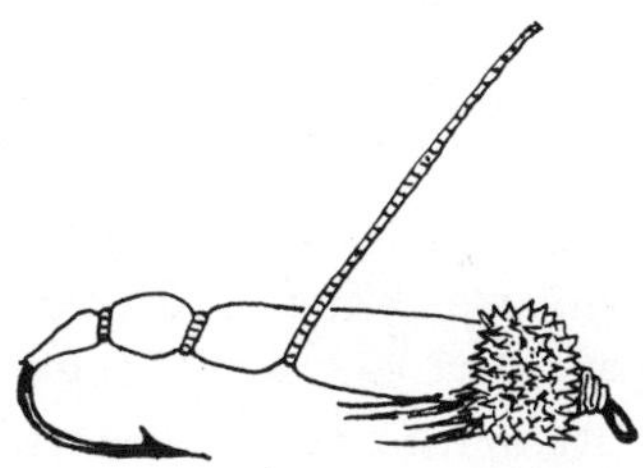

Surface texture is a related issue. Some of the insect forms I like to suggest with this technique have smooth, skinlike exteriors, and we'll talk more about several applications of this sort a little later.

Finally, you can easily indent the body with ribbing to get true silhouette segmentation (as opposed to the shade banding that comes with overlapping the skin). The soft underbody is ideal for this purpose. As with some of the other qualities I've mentioned, this one may be more pleasing to the tyer than important to the trout, but isn't fly tying largely an amusement anyway?

Before looking at other trashliner patterns, let's review the caterpillar in more detail. Whether or not you'll want a long-shanked hook will depend on the insect in question, and, of course, most caterpillars are elongate, so a 3X or longer hook would be appropriate. If you wish to weight the hook, this is the logical time, and in any case it hardly requires much skill to build a body to the desired fullness and taper. I've used polypro, cotton and wool yarn, fur dubbing, foam rubber, and kite string—anything that's somewhat soft and of the correct color for my purpose. Next, trim off a symmetrical smooth-sided strip of liner, 4 to 8 inches in length and one-eighth to one-quarter inch in width. The smaller and shorter the hook, the shorter and narrower the strip. Taper both ends as shown; otherwise you'll get flares at the bend where you tie in and at the head when you tie off. It may be difficult to judge just where the tie-off end of the liner strip is going to be when you've reached the fly's head; however, you can just hang the plastic ribbon from the head of the still-unfinished fly by the weight of a hackle pliers and trim to a taper right on location.

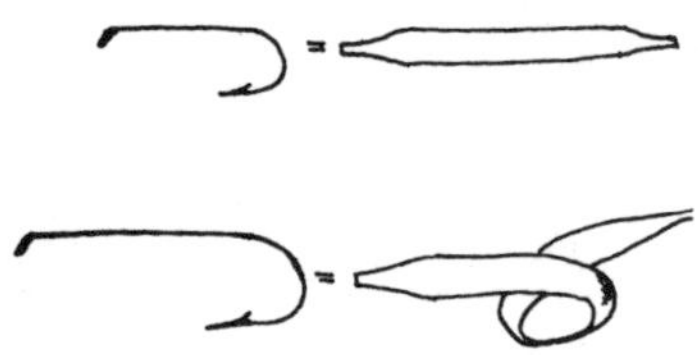

Working thread should generally blend with the body unless you want it to stand out as ribbing. I always rib down the skin with wide-spaced, even spirals of working thread, beginning at the bend and ending at the eye. Naturally, the rib should be quite snug if not tight (as when you're after deep segmentation). A lazy trick that I like is to use just one thread for the whole operation. I tie in the liner strip at the bend, wrap on the skin, letting the end hang by the pliers so that I have two free hands, rib, tie off the liner, add hackle if desired, and whip-finish, all with the working thread. I used soft webby olive hackle to imitate the first caterpillar's head parts and black ostrich herl for the business end of the eager aspen eater. A light coating of head cement over the skin is optional; the fly will hold together about as well without it.

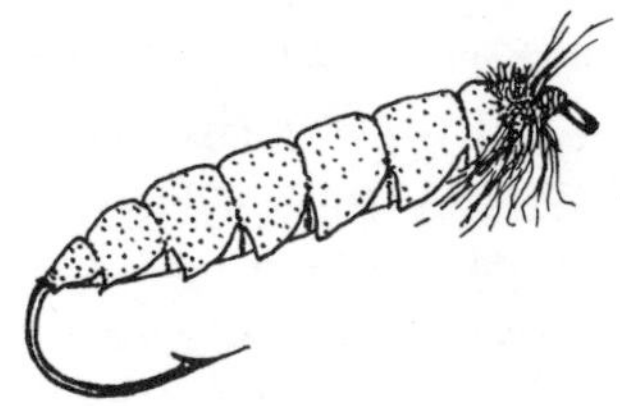

Getting back to that prototype milky-green caterpillar, I found that its very underside was white, not green, so I ended up using a *cape* of plastic rather than a full spiraled wrap. It covered the top and sides of the fly, allowing the underbody to show beneath, and where the ribbing thread crossed the free edges of the cape, the plastic flared in a manner that simulated the caterpillar's stubby legs beautifully. For once it all worked out the first time I tried it. I had an excellent imitation as to size, shape, color (top and bottom), consistency, and even surface texture. I'ts a wonder the thing didn't turn into a moth or a butterfly!

After the olive-drab caterpillar reminded me of this tying technique's seeming potential, I went off on what might be described as a "worm kick." I imitated leaf rollers, inchworms, wormlike larvae, and even (blushingly) earthworms. One fly-fisher friend said I should be ashamed, and only partly in jest, I think. Still and all, earthworms are no different from other land dwellers that get into the water, and the really do get in after storms and during the runoff, as all bait fishermen know. So why not tie an artificial "garden hackle" too? No one gets upset over imitations of ants, beetles, crickets, and grasshoppers, and the last three are also used as bait. Speaking of getting upset, I'm going to spend a couple of pages on the topic of weighted wet flies, for this is virtually an inflammatory issue in the minds of some anglers, and most of the trashliner patterns need a little added ballast if they are to be most effective.

I've never understood why feelings should run so strongly over the propriety of adding heft to a hook, but to certain fellow fly fishers the practice amounts to an outright desecration; it's like muddying the Madison or dumping detergent into the Deschutes. I got into the habit of hook weighting quite innocently and without really worrying about whether it was wrong because I found that weighted wets helped me catch more fish—

no more to it than that. Still, the charge against the weighted hook is not an unbelievable one. It's claimed that wrapping lead wire beneath the body materials causes a fly to sink through currents in an insolent and unnatural fashion rather than being carried by them as a real creature would be wafted along by the watery breezes.

Now, there's no question that it's possible to add so much weight that you essentially end up with a decorated sinker armed with a hook, but that's not what we are talking about. Let's start by comparing the actual weights of ordinary spinning lures with these weighted flies, converting ounces to grams as is the modern practice. A lure of the commonest size, say a 2-inch-long spoon, will run almost a third of an ounce or 8-plus grams, whereas a really tiny "ultralight" lure might go a sixteenth of an ounce or 1.75 grams. In comparison, the heaviest flies I tie weigh less than 1 gram, and these are big fellows built on No. 4 and No. 6 long-shanked hooks. The caterpillar copies I've described approximated 0.5 gram, so it would take three or four of these artificials to balance the scales against just one smallish spinning lure. As you can see, a weighted wet fly need not be equated with a lead sinker.

Let me ask a two-part question that I'll bet has never crossed your mind (at least I hope not): How much does a worm weigh—and who cares? Actually, I think it can pay to care. A while back, I decided to find out what various underwater food forms really do weigh and began to bring home specific nymphs as well as things such as caterpillars and even minnows to study with a precision laboratory balance. This gave me an opportunity to compare the weight of the real thing with the artificial fly I *had* been using as an imitation. I emphasize "had" because this was a good many years after I began adding ballast to my wet-fly hooks, and in the process I'd worked out a whole set of flies, each of them weighted empirically according to what seemed to work best. Believe it or not, I found a very close correspondence between the weight of a given natural, whether a nymph, a minnow, etc., and my best or favorite copy!

The chart below gives figures for five food forms, natural and the matching artificial, both weighed in water. It's evident that my dragonfly nymph, caterpillars, and crane fly nymph were a little light, but the others were right on the money, and all of the weighted flies came much closer to the natural model than those tied without lead-wire-wrapped hooks.

Insect & Size of Corresponding fly	WEIGHT IN GRAMS: *Natural*	*Fly Weighted*	*Fly Unweighted*
Stone fly nymph, No. 10	0.50	0.55	0.32
Dragonfly nymph, No. 8	0.65	0.58	0.43
Caterpillar, No. 10	0.85	0.76	0.35
Crane fly larva, No. 6 (rock worm)	0.91	0.80	0.47
Leaf roller, No. 12	0.31	0.30	0.16

(Unweighted flies were tied on the same hooks and with the same materials as the weighted ones.)

I described my reasons for preferring weighted flies to split shot or sinking-tip lines as a means of getting a natural presentation out of a wet fly in *Flyfishing the Rockies*. However, it's hardly fair to ask the reader to dig up a copy, so I'll summarize briefly:

1. I submit that appropriate weighting enhances rather than detracts from the natural behavior of the fly in the water. The fly will respond to currents in the manner of a real insect, or minnow in the case of streamers. In fact, it's the unweighted artificial that acts "funny."
2. Sinking lines need both time and float space themselves in order to get deep enough to influence the fly. On short floats, as in fast water with small pockets, there just isn't this kind of time or room. A weighted fly doesn't need any help whatever from the line.
3. Not only is split shot a nuisance, it's the shot that sinks rather than the fly. Since shot and fly need to be close together on the leader, the poor fly has to drag an anchor around behind it when there's really no need.

Understand, I'm not claiming that all wet flies need to be weighted. That would be self-defeating in the event that you wished to present your fly just below the surface, as in imitating emerging mayflies. The same is true of tiny wet patterns or flies that will be used in lakes or beaver dams where there is no current. Beyond this, I keep somewhat differently weighted flies in each pattern in separate sections of my fly book to fit particular combinations of current, depth, and float distance. I've outlined a scheme below that ought to give you an artificial that weighs and behaves about as a natural insect or minnow of the same size should. Of course, formula is not exact, although it's about average in my own tying.

In this chart I've rounded the weight figures off, for it's obvious that the final weight will vary with hook caliber and shank length and also with the kind of hook and quantity and nature of other tying materials.

Hook Size	*Caliber of Lead Wire and How Applied*		*Weight of Finished Fly in Grams*
18 & 16	*Small* as a single strip or none at all		0.05 to 0.15
14 & 12	*Small,* short wrap, or pair of strips	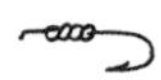	0.20 to 0.35
10 & 8	*Small,* long wrap, or *Medium,* short wrap	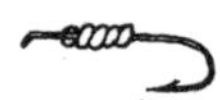	0.40 to 0.60
6 & 4	*Medium,* long wrap		0.65 to 0.90

I've not gotten around to answering the question about a worm's weight, for the notion of a fly that simulates an earthworm is clearly silly. Indeed, my first earthworm flies were tied with tongue in cheek. I had been invited to speak on pattern invention to the local chapter of Trout Unlimited and was looking for a humorous closing topic, as after-dinner speakers should. With this in mind, I tied up several worms and photographed them in color for my talk. Well, they got a lot of attention and looked so good on the screen I decided to really take them fishing. Clearly, the best way to test my fanciful fly was to match it against a real worm (something I hadn't attempted since age nine).

For this grand experiment, I picked a high meadow on the headwaters of the Gunnison River in Colorado where there are more trout than there are good things for them to eat. Stopping at a tackle and bait shop in Almont, I recall a distinct feeling of uneasiness tinged with guilt as I selected a card of bait hooks and a carton of "Worms that Wiggle" (brand name). But I got away with it—no one saw me—and I was soon drifting the writhing annelids into deep pockets on the convexities of meanders along this mountain brook. The water was clear, normal for midsummer and perhaps not ideal for bait, but I managed to catch eleven the first hour. Then my worm fly absolutely delighted me by taking twenty in the next sixty minutes! Back to the real thing, and ten trout were released during hour number three; could the artificial keep pace? It did, to the tune of nineteen and a game-winning total of thirty nine to twenty one, a real upset as far as I was concerned. The major difference between the combatants was due to large amounts of time lost putting fresh meat on the bait hook whereas I used just two flies that entire afternoon.

I decided not to get too pumped up about the whole thing, though, since I had read that a worm really comes into its own in high muddy water when those good juices leak from its poor punctured body and attract fish that might otherwise not find the bait. Such a situation was soon provided by an August deluge that turned New Mexico's Chama River into cocoa. Taken by surprise, I had no live crawlers along; however, three adept companions provided plenty of competition for the worm fly with both nymphs and streamers. And when it was all over, my unwholesome invention was in a tie for first place with a white-winged marabou streamer, having edged a brown Wooly Worm and run several well-known nymphs right out of the stadium. So the pattern appears to have promise, although this all happened recently and I can't comment further.

Whether or not the Worm Fly proves out to be an asset, it taught me one extremely valuable lesson about tying all the trashliner patterns. A worm is obviously an elongate creature, but if you tie one on a 6X long hook, you end up with a ridiculously rigid creation that more resembles a toy submarine then anything alive. So I quite literally learned to put a wrinkle into the fly by bending the shaft at several points and in several directions with needlenose pliers. This does not weaken the hook, and as long as the point and eye are in line when you're through, there's no loss of hooking potential. The bending enhances the fly's resemblance to a worm, but more important by far, the thing will now tumble and twist as it drifts along in a most inter-

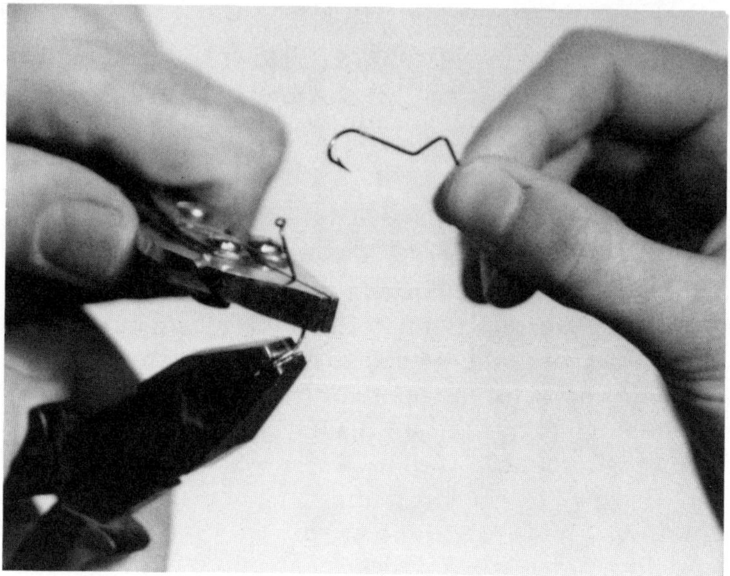

To make a Trashliner look like a wriggly caterpillar and not a miniature submarine, the hook must first be wrinkled. In this photo, one hook has been suitably bent and another is being bent by means of two pliers. To prevent loss of hooking potential, the bend must keep the point and eye in line.

esting wormlike manner due to its contorted configuration. I'm absolutely positive this hook bending makes the fly better, and I now put at least a simple mid-shaft bend into any and all trashliner patterns that have elongate bodies, worm copies and otherwise. The crooked hook is not just a superficial frill!

To finish with the worm, I've found that the small, lively kind that get washed into streams weigh in at about 0.60 grams (as opposed to the fearsome night crawler). After weighting the hook, I cover it with either beige or pink yarn. The skin is the opposite color (pink over beige or beige over pink), and I use a neutral thread color for ribbing. If you really get carried away, you can gild the lily by adding an egg case together with a simulation of a worm's most private parts, as shown in the illustrations. Try the fly after summer rains have roiled a stream with undercut grassy banks, but don't let anyone catch you, either a flyfisher or a wormer. The former will likely be disgusted and the latter infuriated. The bait boys just don't think worm flies are fair, and maybe they have a point. For instance, is it legal and morally ethical to fish the pattern in "fly only" water? I'm not going to say whether I've done this or not.

One of my favorite trashliners is an imitation of free-swimming caddis larvae. Ordinarily we think of shaggy, fur-dubbed bodies for caddis copies, since Trichoptera are covered by hairy filaments. However, on the larvae they're so delicate as to be nearly invisible, and you can get an excellent color match with this method plus that illusion of depth. I've had awfully good luck with pale larval imitations tied on 2X long No. 14 hooks in either pale green or dingy "ring around the collar" white. A thin body base of pale color allows the hook shaft to show through vaguely, like the vein in a shrimp, and for this cotton chenille is ideal because it becomes quite translucent when wet. Just cover white or light-green chenille with pale-gray liner, rib, and palmer-hackle the front end with a small dun feather to simulate legs and antennae. (In the past year or so, several companies have come out with caddis copies utilizing clear latex covering a body base on a bent hook.)

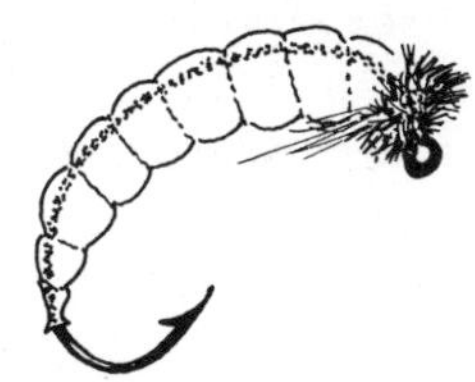

The beauty of this trashliner system lies in the opportunities it offers for innovation in terms of color combinations and always with the option of reversing the base and skin components. The relative transparency of the skin is another useful variable, as is single versus double layering of the plastic. Crane fly larvae are ideal subjects for the trashliner technique, and think of what you can do with freshwater shrimp imitations. Those difficult shades of red, orange, yellow, green, and chartreuse are no longer a challenge; there's really just no limit.

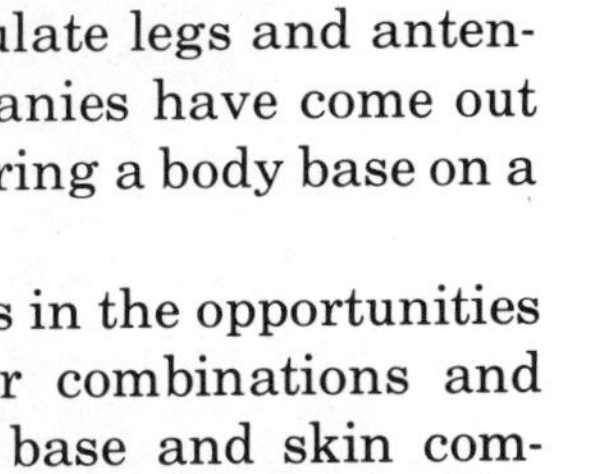

All the patterns we've talked about so far have been "wet." However, you can build realistic dry-fly bodies in this way too. Say the natural is a grayish mayfly dun. All it takes is a thin base of nonabsorbent polypro covered by a skin of clear gray plastic ribbed with thread for stability and segmentation. This construction is particularly suited for the scantily hackled dry flies of the currently popular type that purposely display the body right in the surface film for appraisal by especially critical trout (as during emergence of mayfly duns in still, flat water). The inherent lightness of trashliners can be exploited for imitations of semifloating insects, too. For example, you can do a nice job of imitating a leaf roller or small caterpillar with a slightly bent 2X long unweighted dry-fly hook. The liner skin will discourage water from entering, while the polypro base makes a poor sponge to begin with, so the tiny worm will actually float on a relatively quiet surface.

The trashliner concept is not entirely new. Multiple layers of cellophane wrapped around the hook shaft were used as a means of building bodies at least forty years ago, and cleverly tinted

A student in the author's flyfishing class is shown tying one of his bent-bodied Trashliners. She's in the process of wrapping on the caterpillar's body skin.

latex has been featured in a popular series of soft-bodied nymphs I first tried back in the '50s. More recently, heavy oval nylon that can be colored with a felt-tip pen is being employed as a body material. Latex and nylon are more durable than liner skins; my flies sometimes start to tatter after being toothed by a dozen or so trout (not an altogether unhappy situation). However, the trash-liners are so easy and inexpensive to tie that I usually just strip them down to the body base or, if need be, all the way to the lead wrap with a razor blade and retie them.

I favor the plastic skins over nylon or latex because of the wide-open opportunities for improvisation they offer. I'm always on the lookout for new colors. It's a challenge, and you can't be shy about it either. People are often curious when they see you snipping away at waste receptacles, and it's a rare janitor who will believe your story. It doesn't require a lot of material, though; a 6-inch square will go a long way, and a whole assortment of colors packs up neatly in your tying box. Nor are liners the only source of suitable plastic sheeting. Shopping bags, bread sacks, and lingerie containers can be most intriguing. The most attractive tan/pink I've found came from a pantyhose package I picked out of a trash container. (I think they call it "nude.") Of course, you'll want to develop your own sources.

This chapter makes something of a hypocrite out of me, since the trashliner patterns tend to be more imitative than impressionistic and I started the book by beating the drum for impressionism. I don't mind, though. It's OK. I'm happy to work both sides of the street now and then when it suits my purpose. After all, it's not a moral issue.

CHAPTER 7

The Little Hopper

When I decided to write this book, I began by listing all of my inventions and modifications of other people's patterns on a large pad of paper. It was such an encyclopedic collection that a multiple-volume set would have been required to include them all, and I could neither generate much enthusiasm for writing such a tome nor could I imagine anyone wanting to read it. Eventually, a nice blend of natural sloth combined with a partially developed sense of honesty made things much easier for me. Once I'd culled out the relatively ineffective and mediocre members from this vast swarm of bogus bugs, there weren't that many left to write about, and I felt that the dozen or so survivors actually had something to offer. At that time the Little Hopper was not among this select group, even though I had a grasshopper pattern that I liked very much. My hopper got excluded from the elite corps because, to put it bluntly, there's nothing quite so trite as another grasshopper imitation. Everyone ties hoppers, professionals and beginners alike. They have at least as much appeal to tyers as they have to trout, probably more. I suspect psychologically it makes primitive sense that fish prefer bigger bugs, particularly big fish. Compared to minuscule midges, modest mayflies, and compact caddises, a hopper must represent a veritable banquet floating by. Therefore, it's not surprising that there are so many recipes for artificial hoppers, and what's more, the better-known patterns I've tried have generally done the job pretty well. So what's to contribute? I asked myself. It seemed that the fly tying world had been intent on inventing a better hopper for a long time, the way the rest of humanity concentrates on the better mousetrap, and I could see no point in adding to the confusion.

Having painted myself into a tight corner, I'm now going to try to convince you that the Little Hopper got onto the varsity on the basis of being *little* rather than because of unusual construction materials or principles. Have you ever stopped to think that the larger the insect you are trying to imitate, the more difficult the task becomes? Big flies inevitably mean coarse construction detail, and as detail enlarges, the trout get a progressively better look at the counterfeit you're attempting to pass. In a very real sense, you are magnifying your masquerade. Conversely, as imitations become smaller, detail is miniaturized, and this means the fish will get a poorer view of the facsimile. Perhaps the trout perceive smaller flies as a whole, impressionistically, rather than dissecting out individual components for examination. Legs on hopper imitations are a case in point. Some patterns employ stiff, heavy fibers to simulate the trailing hind limbs, and it's hard to see how these can be too realistic in the scale presented by a fly tied on a No. 8, 10, or 12 hook, the usual grasshopper sizes. These ersatz legs are usually a symmetrical pair, too, and this presents a problem. Nature obviously likes bilateral symmetry, but hoppers are more than likely to float with their legs akimbo, one alternately extended and one drawn up as they struggle to escape the strange watery environment in which they find themselves. Thus, I think it's much easier to blur limb detail

in a smaller imitation. We'll come back to this presently.

I discovered the delights of doll-size grasshopper flies through a weakness I have for ferreting out the secrets of little streams. It's sometimes an embarrassment, but I have a thing about peeking under the willow-fringe skirts of tiny brooks and rills while more manly anglers are showing their mettle on mighty rivers or meeting the challenge of exquisitely difficult waters of the ilk of the Letort, Yellow Breeches, or Armstrong's Spring Creek. Now, if you're working your way up a closetlike canyon where the flow is only a few yards across, individual currents are measured in inches and a No. 10 fly looks like an ocean liner in the locks of the Panama Canal. So in situations like this I'd sometimes try a No. 14 grasshopper, tiny in my estimation, and when the undersized flies were successful. I went on to experiment with No. 16s and eventually, even with No. 18 hoppers. I grew so fond of the little flies that this project logically spread to larger waters, and on the basis of these studies, I would like to share a set of findings that essentially constitute my "defense" for small grasshopper imitations:

1. Real hoppers come in all sizes. Some are pygmy species that never grow large, while others progress from tiny nymph forms into sizable adults in the late summer and fall. So there are small grasshoppers in the nature.
2. Strike:hook ratios are often more favorable with smaller than with larger grasshopper imitations.
3. Trout, including large ones, don't demand hulking grasshopper copies; they'll rise every bit as readily to dainty ones, probably more so.
4. The combination of slow glassy-surfaced water and sophisticated hard-fished trout usually constitutes an angler's stiffest challenge. Happily, this is precisely where and when the Little Hopper does its best work.
5. Even on swift, rough water the fly's thick head and body make it fairly easy to see, right on down to No. 18s.
6. Although not a high-floater, the pattern stubbornly refuses to go all the way under and requires far less drying and doping than a standard dry fly.
7. The Little Hopper is a superb fly to cast in the wind and has an interesting way of landing on the water, as I'll explain shortly.

My first grasshopper was a copy of the still-popular Joe's Hopper. I replaced the yellow body with one of extremely shaggy dubbed squirrel hair in a reddish brown ribbed with bright-red floss. The gaudy color touch was in deference to the flashy patches on the hopper's wings and femurs, but otherwise I used the same turkey wings and brown/grizzly hackle mix as in a Joe's Hopper. It was a good fly, and I was quite content with it until I read Vincent Marinaro's fine chapter on grasshopper imitation in *A Modern Dry Fly Code.* He preferred a large feather quill for his artificial, since a hopper's outline is smooth-surfaced and blocky, not ragged-edged and hairy. My fly's body most resembled a bottle brush, so when the Humpy patterns became well known, I switched to deer hair bound on in a bundle parallel to the hook shank. This model looked better to me, but it wasn't long before the Muddler Minnow was invented. Not a bad grasshopper imitation in its own right, the Muddler's fat head of clipped dear hair added to the pattern's streamlined figure, so I joined a great many other tyers in borrowing this feature for our grasshopper imitations. Although it was an excellent floater, I still wasn't happy, for now the big turkey wing began to bother me. The heavy feathers had a way of making the fly lopsided or unbalanced when it was cast so that it tended to spin, twisting the tippet and causing problems with wind knots as well as with precise placement. And there was just no way of keeping the wing from splitting into a disorderly array of spikes, admitting that the sticklike splintered outline perhaps simulated flailing legs and that the fly was a good fish catcher. Still, a decent artificial needs something to suggest the powerful rear legs and wings. Then, as I began to pay more attention to how a real grasshopper looks afloat, it became apparent that they usually have their wings folded back flat to their sides. Thus, it seemed reasonable to drastically reduce the fly's wing to a couple of trailing fibers on either side of the body and flaring slightly to the

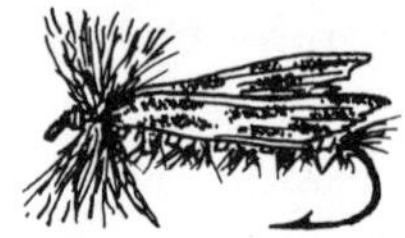

QUILL

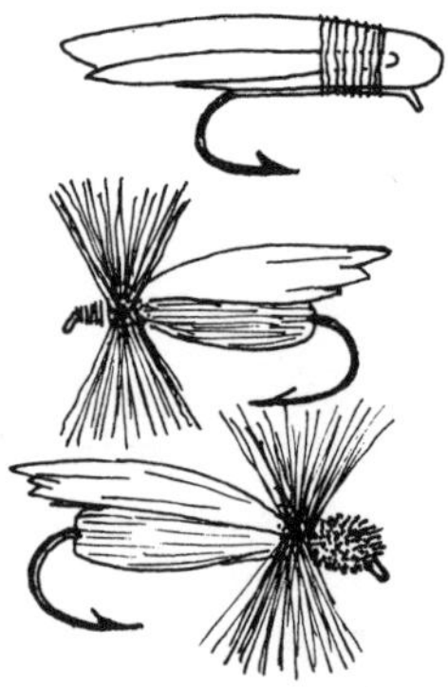

rear (like legs). In accordance with my original remarks about rigid bilateral symmetry, I determined to use fairly thin pliable fibers, either single hairs (moose, for instance) or a pair of fibers from a large quill feather.

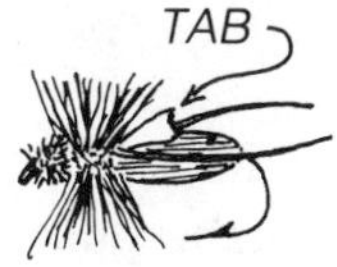

As a last adjustment I got rid of the showy barber-pole rib. It was just too much color on a small hook, and besides, the floss didn't hold well on the slick deer-hair body. Therefore, the floss was replaced with brown working thread, cinched down tight, and for color, I left a tuft of floss protruding just behind the hackle, as shown.

At this point I'd like to digress a bit on the Little Hopper's casting qualities. I think they are important, so if the point seems lost for a minute, please hang on. Those of you who have ridden the picturesque narrow-gauge railroad that straddles the New Mexico/Colorado line will remember the pretty Pinos River. Descending eastward from the summit of cloud-capped Cumbres Pass, the railbed follows the meandering stream for many miles, high above grassy meadows, before the Pinos plunges into the Stygian depths of forbidding Toltec Gorge. The view of the river is sufficiently seductive that few fly fishers remember much about the historical railroading aspect of the trip. I know I would have had to have been physically restrained from jumping train on my first ride had I not been traveling without tackle at the time. Far wider than deep and clear as crystal, the Pinos contains some nice although unbelievably suspicious browns. Indeed, the trout are often offended when an angler approaches to within 60 feet, and a 5X tippet looks about as subtle as a water snake gliding along the surface. But that's not the worst of it; there's the wind. It has been my experience that warmer air rising from lower valleys toward the summit of a mountain pass usually creates an upstream breeze, and this is clearly what should happen on the Pinos. Nature has it the other way around; an upstream-facing flycaster gets it right in the face most of the time, and at a rate of 10 to 40 knots too (and I mean 40). I'm sure you see the problem.

It happens that a fisherman working the Pinos is occasionally in view of a whole trainload of folks, up to several hundred, as the puffing locomotive pulls into view, draws even, and gradually chugs off into the distance. Now, as a teacher I'm used to lecturing in front of an audience, but a solo fly fishing performance is another matter. It's very much a matter of being on public display. And you get everyone's closest attention, too, because the passengers won't have seen another large mammal aside from a few livestock and an occasional deer for many, many miles. You can almost hear the camera shutters click and faint requests to "wave" waft down from the open observation cars far above, so there's real pressure to produce. Having given this some thought, I've come up with the following idealized script: Wait until the locomotive is just about opposite and then hook a large brown, preferably one that will jump. While the fish is airborne, hold him against the real, thus freeing your left hand. Turn toward the train, remove your hat, and bow; that's how to do it.

One day I had the whole thing planned. The train was coming and I'd spotted a good fish rising steadily just about 40 feet upstream under a protective grassy bank. They'd been taking one of my earlier grasshoppers in a size 10 fairly well, and while the cast would be tricky, I was pretty confident. I waited, working out line via a low sidearm false cast and crouching slightly (looks more professional). I'd need to shoot just a few feet of line and the breeze should drop the fly just above the brown in a gentle curve with some slack in the terminal leader. Now it was time; I thought I heard "Catch one!" between chugs and I let her rip.

Well, it was a bad scene. As nearly as I can figure, it must have been a freak zephyr or dust devil, for the line shot out, on target, and then suddenly rose straight into the air like a cobra from a basket. The snake instantly developed coils which proceeded to shoot back at me with such tremendous velocity that I was helplessly enmeshed in a mess of loops, and in the process of trying to dodge this flying net, I stepped backward, slipped, and sat down in 18 inches of icy water.

I didn't wave, and I hope the children aboard didn't hear what I said, either. I think there's a moral or fable about this sort of experience, but you're probably wondering what any of this has to do with the Little Hopper.

Although there's no way to prove it, subsequent experiences on the Pinos and equally windy waters lead me to believe that the whole disaster could have been avoided had I been using a Little Hopper instead of that big artificial. The grasshopper I had on was one of the old kind with a generous hackle and a large pair of feather wings that probably functioned as a good stiff sail; the fly acted like a kite on the end of a string. Perhaps this is an exaggeration, and it's likely my fly would have gone somewhat awry regardless, but the Little Hopper is truly great in wind.

The reasons are as follows: Its surface area is small, since there's no true wing and I use hackle about one size small for the hook. The fly's contours are smooth and bulletlike because of the shaped, clipped hair head and body. Further, the head and body will trap or absorb several droplets of water, making the Little Hopper slightly heavier than a standard dry fly of the same size. Add these qualities together and you have a projectile pattern, a fly that will go right through the gale and toward its intended destination, relatively impervious to capricious breezes. No doubt the fly is only one component of the tackle chain, and the rod, line, and leader have to be up to the task as well. Still, when I did my trick in front of the train, I was armed with a rig that was chosen especially with wind in mind, so I really believe it was the fluffy fly that did me in.

If it seems I've overemphasized the casting behavior of the Little Hopper, I'd like to point out two related phenomena that are probably even more vital. Because of its bulletlike flight through the air, the fly has considerable velocity as the forward cast unfolds such that it pulls the leader out quite straight on the water. This will likely happen even if you've gone out of your way to cast a slack line, and for the same reason, it's also likely that the fly will hit the water hard, making a little splash. In fact, on a quiet day in still water there can be an audible "plop."

Although we tie our hoppers differently, Mr. Marinaro described both the straight leader and the hard sitdown in his *A Modern Dry Fly Code* over twenty five years ago. He explained why these peculiar physical attributes of artificial grasshoppers are not necessarily negative features. In the first place, a grasshopper leaping from the bank doesn't make an Olympic diver's ideal entry, so the "spat" (as Mr. Marinaro called it) is actually realistic. Second, a straight leader often leads to almost immediate drag, since there are no squiggles to damp out differences in current velocity. In the case of a grasshopper dry fly this sort of early drag can be another plus, because the real insects clearly don't feel comfortable in the water and struggle strenuously to change their situation. Muscular creatures, they may quite literally row their way across one feeding lane into another, countercurrent if you will.

As it happened, I read all of this with interest many years before the first prototypes of the Little Hopper were tied, although I never tested Mr. Marinaro's observations for myself. To me, drag was the dry-fly fisherman's *bete noire,* something to be avoided at all cost. However, when the Little Hopper was developed, I had no choice, since you get a ruler-straight leader and fly splash whether you want them or not. I should have paid better attention to Mr. Marinaro, for my fishing log is just full of confirmatory annotations. Time and again, in the process of recording a day's events, I have been impressed enough to comment that the trout either hit the hopper instantly upon its arrival or took it dragging or some combination of the two, and these facts were jotted down solely because they surprised me somewhat. I'm no longer surprised.

Before we leave, there is one last aspect of the physical behavior of this fly I'd like to mention that makes for especially exciting fishing. Because the Hopper makes a disturbance on impact, you can usually pick it up right away visually, and since there will be little or no slack in the line and leader, you can also set the hook immediately in the event of a simultaneous strike. It's like the "bang-bang" play baseball announcers love when a sharply hit ball gets caught before the ring of the bat can fade. There are lots of bang-bang days with the Little Hopper!

Wind or no wind, slick-surfaced meadow streams like the Pinos tend to be moody, and when the trout are sulking, it's hard to get them to come up for anything. These are the situations in which I've had particular success with either the Little Hopper or the Black Widow, the subject of the next chapter. Tied on No. 14, 16, or 18 hooks, these compact patterns have virtually retired my big hoppers, for no matter how deep their feeding doldrums, there's always a good chance the trout will wallop these small terrestrial imitations with vigor and sincerity.

I prefer gray-brown deer hair for both abdomen and head, although there are other options as regards hair type and color.

Nature's hoppers come in near-whites, dark grays, yellow, brown, and green, but I've done best with hair right off the deer. You might try lightweight, coarse-fibered antelope hair or finer, tougher elk instead. The Humpy type of body construction in which the fibers are tied on in a parallel bundle is fine, although I've come to like the "hair on end" irresistible style better. It's probably superior for flotation and is certainly less likely to fall apart.

I've noticed that quite a few anglers are timid about attempting hair bodies of the Irresistible type; I certainly used to be. Retrospectively, there is surely no construction in all of fly tying that appears so difficult and yet in reality is so simple. As a former President commented in a different context, the only thing to fear is fear itself, and the first time you try it, the insecurity will just disappear. When I talk about the Little Hopper to a group, there is invariable concern about the propriety of putting both a hair head and abdomen on a smallish hook. Actually there's nothing to it. What you do is to make the head first thing while the hook is still naked. Trim it to final shape, shove it up to the eye, and forget it. The rest of the fly is finished as if the head weren't there. Next the abdomen goes on and is finished off leaving about one-sixth of the hook shaft bare. With a 2X long dry-fly hook such as the Mustad 94831, you'll still have room for the bright floss and thin soft feather-fiber leg tie-ins; they don't need much room at all. I generally use gray-brown fibers from a turkey quill, always thin and pliable. The last step involves tying in two hackle feathers, one red-brown and one grizzly, both about a size small for the hook. Advance the working thread to the

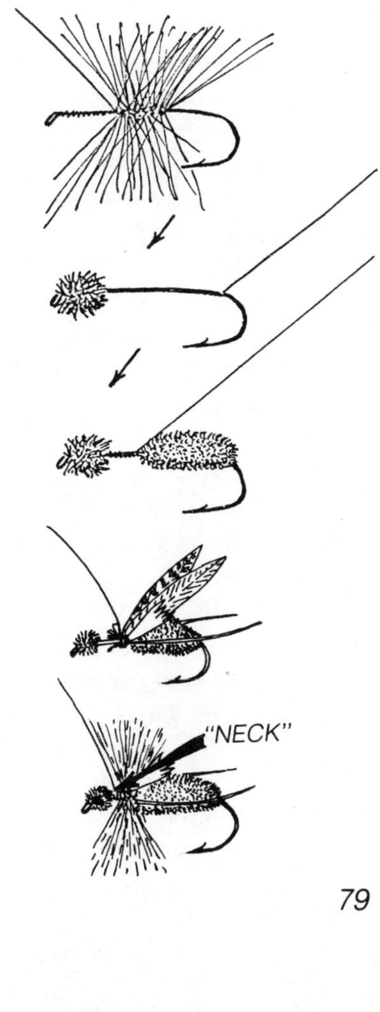

very back of the completed head, wrap in the hackle, and whip-finish as a tie-off. This leaves a short neck just behind the head and in front of the hackle. One friend likes to add eyes with drops of acrylic paint, but this is too fussy for me; there is already a lot of miniaturized detail, and some of his hoppers look cross-eyed!

The kind of troubleshooting assignment the Little Hopper likes best is illustrated by the following anecdote. We were on the North Platte River in Wyoming about a mile or so above the Colorado line in what is called Northgate Canyon. Here the river alternates between long shallow flats with very little current and shorter, steeply pitched rapids. The pondlike strips are seldom productive, although they contain a few large fish that are exceptionally easy to spook, and we usually walk around the flats to get at more detailed water. Just as the Platte has low yield areas there are also off periods as to trout activity. My son Charlie (who was fourteen at the time) and I had been futilely flogging the water along opposite banks for some four hours when Charlie came upon a rare hot spot along the edges of a swift current. And from this strip of riffled water he proceeded to extract a whole series of plump rainbows. Charlie did it all with a Golden Adams, and in a manner I thought somewhat condescending he dangled each victim from his leader for my inspection prior to its release. Of course, I was pleased for him; no matter if I hadn't had a strike myself for nearly an hour. It doesn't bother me that much when companions do better than I; it happens a lot, and I usually can blame some peculiar pattern with which I'm experimenting. (By the way, this is a significant bonus for the pattern inventor.) With Charlie it's different, though, for his is a tenacious memory, and you're likely to hear about a particular incident for a long time, surely until the next Christmas, and with sporadic bursts of recall there can be further flashbacks years later.

So I badly needed to catch some fish right away. But now I'd run out of likely looking water along my bank and had reached the lower end of a very long, listless flat that stretched on above and around a bend. I felt I should stay within earshot, since the youngster was crotch-deep in fast water, battling rainbows right and left, and I had some residual parental concern for his safety. Charlie hooked another; I was torn. While walking the trail downriver that morning, I'd noticed this specific piece of still water. A small draw led a rivulet into a marsh along the far bank, where a defunct beaver house protruded into the river, and it looked as if there might be deep water on its downstream side. It was ideal Hopper territory. Charlie's shout announcing number eight in his growing series of triumphs made the decision easy enough, for there was just no way to extract him from that riffle, and I most certainly wasn't going to sit on the bank and watch. Wading across into casting position, I violated an earlier agreement that each of us would stick to his own bank, but this time treaty-breaking paid off. The Little Hopper promptly hooked, and I landed four nice trout and, additionally, turned over the largest fish I've yet to see on the Platte. As it turned out, Charlie was pretty much finished for the day once he'd exhausted his bountiful riffle, whereas the Little Hopper continued its winning ways in deep pockets of still water, eventually tying the score and saving me from a peculiar kind of fate only fathers with sons like Charlie will understand.

CHAPTER 8

The Black Widow

This pattern and the Little Hopper have a great deal in common. Both might be called "damp" rather than dry flies since they float half submerged where the trout see them in and partially beneath the surface film. And like the Hopper, the Widow's mission is to suggest a land dweller that, by misadventure, has fallen in. The patterns are also similar in that I depend upon them as "hole cards" for difficult situations. However, the Widow differs in that it's not intended to copy a specific insect. Instead, it might suggest a spider, cricket, beetle, ant, wasp, bee, or just about any kind of darkish bug. I really don't worry about which or what the trout think it is, because they often will pound the pattern whether or not there is anything similar in the water or in their stomachs; it's like the Maniac Nymph in this regard.

Unfortunately, I'm afraid the preceding paragraph has a pompous ring to it, and this is neither intentional nor supportable. As the illustration shows, the pattern is very simple, and I'm sure most readers will have fished an ant imitation, either as a wet or as a dry fly, that was similar. So the basic construction is both old and productive, and I have only a modest modification to talk about, although I'll say it again: The Widow is a winner!

The usual ant copy offers a three-part imitative ploy which includes size (tied on smallish hooks), color (black or red), and, importantly, silhouette, i.e., wasp-waisted or better yet ant-waisted. There are several easy tying tricks with which one can form a fat abdomen separated by a short gap of bare hook from a second thorax/head lump. Add a few hair or feather fibers protruding from the gap between to simulate legs and you have a truly artful ant. For a wet imitation, I build the body and head lumps with heavy surgical silk suture material or black nylon by just overwrapping until the desired bulk and taper is acquired. The final layer requires finer nylon thread to cinch things down, and for swifter, deeper streams I start with an underbase of lead wire. This gives a nice bulge quickly and the wire is easily covered with thread. Multiple coats of head cement yield a glossy sheen that's really lifelike, too; however, all of this is just a preamble, for the Widow is a surface fly.

I got interested in a floating application of the ant idea through an experience that might truthfully be described as startling. As mentioned earlier, my wife comes from southern Idaho, not far from legendary Silver Creek in the Sun Valley area and, as you might guess, there are expert anglers in her family. Aunts, uncles, and various and sundry other relatives are well versed in the vagaries of sensuous Silver Creek, and so when I first visited Idaho, I lost no time in seeking advice from my new friends. There is no quicker road to success on unfamiliar and challenging water, and knowing this, I listened very carefully and with sincere respect. Nonetheless, what I heard made me frankly skeptical, for it appeared that the local folk preferred to fish demanding

Silver Creek with course-bodied Irresistible-type dry flies in sizes 10 and 12. Now I had no particular prejudice with regard to Irresistibles, but you can't imagine how crude and out of place one looks sitting out on a glassy surface, especially when the water is perfectly clear and the quarry consist of highly sophisticated and well-fed rainbows. To my eye the hair-bodied flies more resembled water shrews or tiny ducklings floating along than insects, and beyond this, the locals didn't pay a whole lot of attention to color either. Some liked a black fly while others chose blue dun, grizzly, or ginger; the only unifying theme seemed to be the fat body.

This was at a time in my career when I could have been described as an entomological snob or "bug bore." You know what I mean—the scientific Latin name bit for family, genus and species. Having a background in biology, I was pretty good at it, too, since I was used to memorizing things of this kind and could enunciate the polysyllabic names with dash and flair. (There's a parallel among wine buffs—knowing how to spell the names and regions and chateaus isn't enough; you've got to have the pronunciation right as well in order to move comfortably in this sort of fast company.) A fatuous fly fisher wouldn't dream of describing a hatch as "some gray things coming off the water." First, he'd identify them as mayflies (Ephemeroptera would be better), and then as subimagos ("duns" is all right, although a bit pedestrian), finishing with the genus and species for a touch of panache. Most proper insect names have a grand sound to them, too. Take, for example, *Siphlonurus occidentalis.* And if some wag suggests that this sounds like an oriental venereal disease, the remark is definitely not funny!

So I stuck with delicate matching patterns, and I did pretty well, too, at least for numbers of fish caught. Size was another matter, though. While I collared an occasional rainbow in the 14-to-16-inch class, the real trophies were all taken by my in-laws on their big-bellied flies. Sure enough, the old saying about beating and joining really is hard to refute, so one summer morning I finally set out with the intention of following the advice I'd received. Aunt Verda had just removed a multipound brown from Loving Creek, one of the big stream's major tributaries, and I was in possession of detailed instructions as to just where and how to proceed, plus an assortment of her favorite patterns.

In contrast to Silver Creek, Loving Creek is far from lovely to look at—small and sluggish of current, with banks occasionally crowded by cattle, while vast floating beds of water weed clog the channel. In fact, there are places where 80 percent of what would ordinarily be water surface is replaced by hideous clots of chartreuse and brown vegetable matter actually drying in the sun. It's a pastoral place; the air reverberates with the hum of insects punctuated by moos, the squawks of blackbirds, and an occasional far-off coyote's bark.

I still couldn't resist trying my usual patterns, although by eleven a.m. I hadn't seen hide nor hair of a fish and was getting downright sleepy in this soporific setting. It seemed as good a time as any to offer them an Irresistible; at least I wouldn't be wasting precious minutes during a hatch. So I tied on a hair-bodied version of a Rio Grande King. Now things were worse. Not only was the Irresistible ignored, the fat fly was just about impossible to cast with any degree of accuracy on a fine tippet. At times it was difficult to hit the water at all, because most of my casts had to be long and the strips of open water were sometimes two feet or less in width. Even when I did manage to get the fly into the water (sounds silly, I'll admit), the corpulent thing would sit there, bobbing foolishly, nothing happening and nothing to do about it.

I yawned. Perhaps a nap? I was moving slothfully upstream, amusing myself with the flycasting target game, when it happened. After several misses, I had managed to land the Irresistible on a thin pencil of slowly moving current near the opposite bank where the fly sat for a good fifteen seconds, unmolested as usual. Then I have a clear recollection of the surrounding mat of weeds beginning to swell and heave upward as a deep pit developed in the water where my fly had been. At that instant a vague, tawny form passed through the vortex of the whirlpool, a thick-bodied thing somewhat reminiscent of a beaver. And there followed an indescribable sound combining qualities of a loud slurp, gurgle, grunt, and percussive thump accompanied by a tidal-wave heaving of the weedbeds from bank to bank, upstream and down, for some distance. As in the case of the float trip I took down the Roaring Fork without a boat, this tale would be a lot better if it had a fishy climax, but one shouldn't lie about such things. My leader came back broken and bare. (I do think the attacker was a brown rather than a beaver, though.)

This encounter went a long way toward knocking me off my high horse of nearly total preoccupation with hatch-matching patterns, and because of my fondness for the wet ant pattern already described, I began to work on a floating version. Copying the two-lump body of the wet fly, I simply took a piece of soft-fibered deer hair of the kind that's so nice for Muddler Minnow heads, dyed it black, and tied the separated parts of the body, Irresistible-style, at either end of the hook shank. After I added a short, skimpy black hackle amidships and trimmed it top and bottom, the fly looked appealingly like an ant. To put it differently, the pattern have a very high "buggy index" rating. (You test for this quality by dropping an experimental fly on the kitchen floor when no one is looking and then watch to see how many family members start to stamp on it.) This favorable index convinced me that the ant should work wonders, and I had hardly forgotten the Monster of Loving Creek, so my new pattern saw plenty of action.

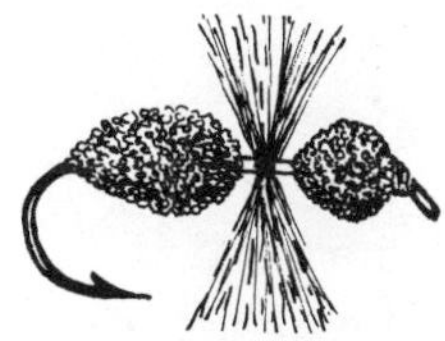

After approximately two seasons the ant had piled up quite a series of adventures, and on this basis I came up with the following observations: (1) The fly had a gratifyingly broad range of appeal to the trout as to time of season and locale, but it was most effective on relatively quiet water and in the middle through the late season. (2) The fly's productivity bore no absolute relationship to the presence of natural ants. (3) At least in my hands the ant performed far better in sizes 14 and smaller. (4) It was not a particularly good floater. (5) The pattern was generally quite difficult to see on the water, partly because of its color and partly because of my preference for small hooks plus a low float profile. So it was a mixed bag of credits and debits.

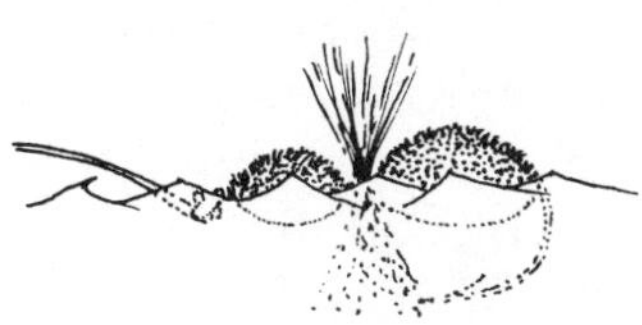

On balance, I found that I didn't particularly enjoy fishing with the ant because of an uncomfortable feeling that I was missing at least half of the rises that my fly solicited. It was a lot like flying blind, and I'd frequently be a good 6 feet off in my estimate as to where the ant should be when a strike came along or when I was in the process of picking up for a new cast. A nymph fisherman who presents his artificial dead drift is used to this kind of uncertainty; it's part of the game, and besides, he always has the leader to watch. When you lose visual contact with a dry fly, though, it's blind man's bluff. I thought that if I could get the ant to float higher in the water, this should help, since very dark flies tend to stand out pretty well if they are up on the surface.

It was clear that the effete mid-shaft hackle wasn't sufficient to hold up both the head and abdomen after they had absorbed a little water, as Irresistible bodies are prone to do. So accordingly, I decided to lighten the body. This I accomplished by replacing the hair head with beefed-up hackle, while maintaining the gap of bare hook between the abdomen and hackle to give the desired effect of segmentation. I also tied the hackle quite thick and used feathers one size smaller than normal for the hook in question, just as one does in tying a dry fly for especially rough water. Looking up at the floating fly through the bottom of a glass dish, I found that the short heavy hackle further emphasizes segmentation, and on the stream the float was clearly improved, although the ant still wallowed in the surface film, antlike. At this point I felt the pattern was improved, although it was still hard to see and I pondered the addition of a hair or other wing. Of course, if I went this route, I'd essentially have a small but otherwise ordinary Irresistible, and I suspected that the wing butt would blot out the wasp-waist segmentation I liked so much. (I tried it and it did.)

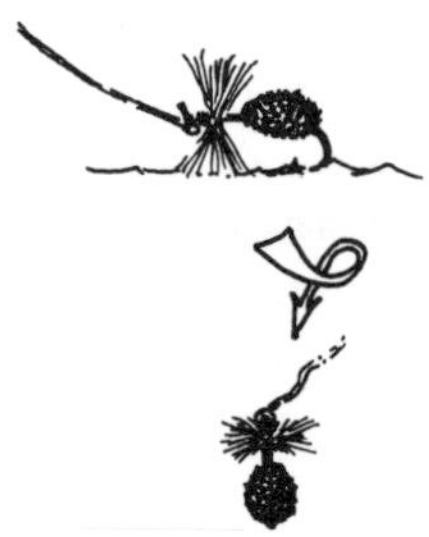

SUBSURFACE

Having run out of ideas, I had pretty much resigned myself to fishing blind most of the time when this pattern was on the leader until I stumbled onto a beautifully remedial solution. It all came about as the result of a fortuitous combination of good fishing and bad lighting, meaning that the trout were taking the ant like hotcakes at a time when I was having even more than my usual difficulties finding the fly, an equation that spells extreme frustration. And I was, in fact, to the point of trying something silly.

When fishing a nymph, dead-drift, under conditions of particularly poor visibility, I occasionally tie a thin, 2-inch-long strand of light-colored polypro yarn to the leader about 30 inches above the fly via an overhand knot so that a leader splice knot keeps the yarn from sliding out to the fly. It's an old trick; others use bits of bright cork or colored tape as aids in leader watching while nymphing. I always carry a few strands of polypro in my leader box, and on this day I selected a tuft of yellow yarn, wondering how badly it would frighten the risers, for I'd never thought of the ploy for finding a dry fly. Fortunately, this didn't turn out to be an issue, since the trout paid the floating fluff of yarn no attention whatever. However, my problem also remained unsolved, for the strike indicator gave me a positive signal only when a trout would hook himself and drag everything under, and I hardly needed help with those guys. As I removed the polypro after this unsuccessful venture, I decided to see what would happen if I caught the stuff right in the coils of the cinch knot that joined the fly to the leader. Sure enough, the fleck of color virtually doubled the fly's visibility, and, more surprisingly, the fish weren't frightened in the least. As a result, I had a most prosperous and happy afternoon, and because the yellow-against-black contrast reminded me of a black widow spider, the pattern was so christened.

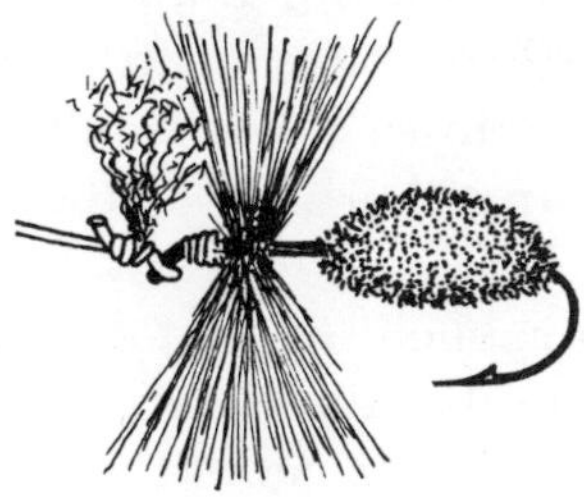

Obviously, I didn't want to have to fumble with tags of polypro every time I changed flies, and I found that the soft, compressible material could be tied in *under* the hackle so that the tie-in was buried, allowing the polypro to flare to the rear on top of the shaft without blurring the wasp-waist silhouette. Note that the tab of yarn is a single tuft, about half as long as the hackle fibers, and does not constitute a true wing. I prefer to think of it as an important "visual aid" for the angler and as an inconspicuous addition from the trout's viewpoint (quite literally). I have repeatedly looked for evidence that the bright color dab might be a deterrent to trout acceptance by first fishing with an intact Widow and then clipping off the polypro from the same fly and offering it plain. As far as I can tell, there's no difference whatever apart from the not inconsiderable advantage and color-tabbed Widow gains in visibility.

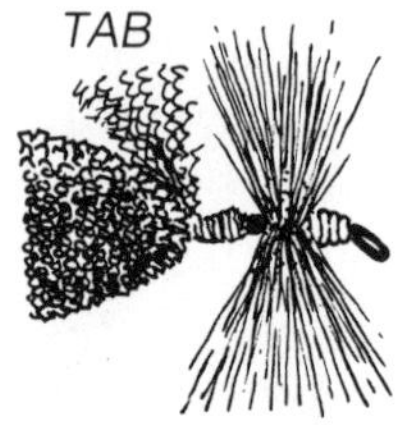

The Black Widow came along some time before the Little Hopper, and the Hopper's stubby pair of color dabs behind the hackle were borrowed in part with the intention of making the fly easier to find on the water. The patterns are also alike in their physical casting properties. The Widow, too, is an excellent wind fly because of its globular, slightly absorbent abdomen and short, compact hackle, while the tiny polypro tab folds back neatly during flight. The Widow also tends to strike the water hard, making a little "plop," but, as with the other fly, trout sometimes literally charge it from some distance rather than fleeing the area of the splashdown in dismay.

You might want to try a soft coarse hair such as antelope for the Widow's body because of improved buoyancy. It takes far fewer of the thick fibers to fill up the body than with deer hair, and I've used both. Remember not to be afraid of tying hair bodies on small hooks. I believe there is innate pleasure involved in trimming the unruly ball of hair you start out with into a neat shape. Manicuring the body into a stylish afro may be messy, but it's also just a lot of fun. Nor do you have to stick with a symmetrical football shape. Indeed, it's important to trim more hair from the underside than from the top so as not to occlude the hook gap. A reduced gap makes for poor hooking capacity. One of my boys went so far as to tie a flat-bodied Ladybug with black hair and hackle. He left the underside black and covered the

surface of the dome-shaped top with bright-red acrylic paint dotted with black, and it too passed the kitchen-floor test when my wife tried to sweep it out into the yard where it could help control the aphids on her roses!

As I've said, the Widow can be a valuable ally when it comes to salvaging an otherwise sour situation on the stream. It has probably had its finest hours on slow quiet water, particularly in relationship to meadowy overhanging banks. A typical Widow "save" took place recently on fabled Fish Creek in the Jackson Hole area. It was mid-August and the weather had been unseasonably wintry for most of a week when we awoke one morning to find the Tetons spray-painted a glistening white right down to the 9,000-foot level. It was a photographer's delight, for the frontal system had finally moved on, leaving us the bright legacy of a bluebird day, and for my own part I happily anticipated a burst of feeding activity. Unfortunately, an angler's anticipations don't necessarily jibe with what actually takes place, and it was soon apparent that the fish simply didn't wish to play.

Giving up on floaters, I caught a few from larger pools on a Muddler Minnow, waiting impatiently for some sign of surface feeding as the pretty morning faded into noon. Eventually three p.m. rolled around, with Fish Creek still silent if not downright sullen. Frustrated, I changed my line and leader, replacing the big streamer with a dainty Widow, for the warming water looked ideal and it seemed reasonable to anticipate some sort of dry-fly action at some point on such a golden day. Nor was my faith misplaced. Cutthroat began to engulf the Widow at once with marvelous confidence and enthusiasm, swimming right down to the bottom with their imagined victim. But something was funny. All of the Widow's customers materialized from the very depths of the stream, the permanent holding water, and never from currents, glides, or riffles, the places trout usually frequent when actively feeding. And there were still no spontaneous rises to be seen. It just didn't compute. Why this kind of secure take in the face of an essentially turned-off stream? These were heavy-bodied fish too, 13 to 15 inches, not foolish dandyprats frolicking about.

In the course of releasing these fine specimens I was impressed that most of them had hard, sausage-shaped stomachs, rigid with food of some sort, and out of curiosity I sacrificed two trout plus a sizable whitefish. The autopsy findings were rather remarkable, even to a pathologist. In each case the stomachs were stretched tight by literally hundreds of flying ants, clearly members of a mating swarm. The reddish-brown drones or males were smaller and lighter in color than the big-bellied queens, whose size, shape, and configuration were closely simulated by the Widow. Surprisingly, a few of the Hymenoptera were still feebly motile, so I had a highly plausible explanation for the Widow's witchery.

I'm sure there was a large measure of luck involved in my success this day. In the first place, it's quite likely that the ant swarm was an isolated event as to both time and place rather than part of a widespread hatch such as those put on by mayflies and other aquatic insects. In other words, I just happened to be in the right place at the right time. My fly was right too, although I'd chosen the Widow on a semi-hunch basis, remembering past performances in somewhat similar settings, I suppose. About halfway through the excitement and before I'd killed a fish I happened to lose my Widow and replaced it with a Royal Humpy, that favorite among all Jackson Hole dry-fly patterns. When the Humpy was ignored, I naturally switched back to a Widow, to immediate resumption of negotiations with the rod-buckling cutthroats. Thus the trout were at least somewhat selective. Supporting the thesis of a localized swarm was the fact that as I worked my way some 200 yards farther upstream, action slowed perceptibly. I imagine I was getting progressively farther from the epicenter of the ant swarm somewhere below me along the bank.

Incidently, the color tab was helpful in keeping track of the Widow even on this sunny day. This was especially important because of the subtle sipping sort of take that was prevalent. On several occasions I lost sight of the Widow despite the tab only to rediscover it a split second too late as it was being disgorged by a thoroughly disgusted cutthroat that had carried its prize way down to the streambed.

Remember that the Little Hopper and Black Widow are similar patterns, since they tend to be most successful in the same situations and settings. However, this doesn't mean that their performances will necessarily be parallel. Many times one "damp fly" will outperform the other, and the trout may even change their preference during the course of a single day on the same stream.

This happened last October on a brief visit to Sheridan, Wyoming. Characteristically I somehow managed to time my arrival with that of a frontal system, and while the storm was neither very prolonged nor severe, it sent air and water temperatures plummeting. This abrupt change came along just as the brownies were beginning their spawning activities, and it all added up to some tough fishing. Snow blanketed the Bighorns, so we turned to lower streams such as provocative Piney Creek, but this time even my favorite patterns failed to impress. It appeared that very rare risers out in the still water away from the main current provided the only game in town, and since these flats were often close to a bank, I tied on a Widow. And through midafternoon it did a good job in the sense that none of us had been catching much whereas the Widow at least managed to woo a finny suitor every half hour or so. But it didn't last. After a span of ninety totally empty minutes, I switched over to the Little Hopper and was pleasantly startled when it collared a fat 14-inch Brownie on the second cast. Gratifyingly, the Hopper kept taking fish steadily, if slowly, until evening.

Now, it may be that the grasshopper copy would have been better from the beginning, there is no way to tell, but this isn't what I'm getting at. The point is that if I'd given up on terrestrial imitations when the Widow became ineffective, I'd likely have drawn a blank for the balance of the day. I say this because two skillful companions, both natives with far better knowledge of the water and its inhabitants than I, came away almost empty-handed that afternoon. So it pays to try both patterns on most occasions.

CHAPTER 9

The Pink Wolf

I know more than a few dry-fly purists who look askance at experimental patterns, for purists also tend to be traditionalists. They prefer the grand old flies, the Quill Gordon, Hendrickson, and Cahills, and I think they are wise in one respect. Old patterns get that way because they *are* good; theirs is a proven track record, so why fuss with possibly fallacious combinations of fur and feathers? Nevertheless, because of an inherent streak of perversity I sometimes go out out of my way to show one of these straight anglers an offbeat fly just to elicit a reaction. Usually they're polite about it. For instance, the Spanish Fly is generally regarded as "amusing" by people of this sort, although you can bet they wouldn't be caught with one on the end of their leader. Not long ago I attempted to stimulate a particularly prissy piscator by handing him a copy of the Pink Wolf. After examining the fly briefly, he glanced up, checking my facial expression for signs of fraud, and returned the proffered gift with a comment I really enjoyed: "It's obscene." Let's see what you think.

By way of an introduction, I'm going to tell two short unpretentious stories, neither of them with much dramatic clout but still real events in the genesis of the Pink Wolf. About a million years ago there was a mountainous volcano near the center of New Mexico that is said to have soared to an altitude of over 20,000 feet, and when it blew its top one day, there were pieces that came down as far away as Kansas. Our Jemez (hay-mez) Mountains west of Santa Fe represent what's left of the volcano's base in the manner of a burned-out firework. The remnant of the crater is an immense grassy plain ringed by minor peaks, and from this high basin pour a number of small streams, like spokes from a hub. In the process of filtering and cutting through the lava debris, these brooks become quite fascinating. Fed by thermal springs, they contain ample food to support a nice population of browns and have developed considerable structural character in the way of pools and riffles, etc. I believe you'd like these streams as much as I do, for it's possible to leave work at noon (surreptitiously or otherwise) and still get in a good solid outing.

It was May, stone fly time on the Guadalupe River above the Jemez Tribal Pueblo, and the nymphs had been active for several weeks. Now the adults were beginning to become plentiful in the willows and along the water, frolicking from rock to rock, inches above the surface. Mature stone flies probably don't go swimming by choice, but they're quite apt to drag their landing gear and wallow around for a second or so. If a trout spots them meanwhile, it's quite likely that a miniaturized version of the sensational killer-shark movie *Jaws* will follow. This particular spring the numbers of stone flies seemed about twice normal, and I thought it would be a good opportunity to test a really careful copy of the mature insect against nonmatching impressionistic patterns.

In this interest I went to considerable trouble to construct a rather meticulous matching pattern. As you know, the real stone fly has a flattened, oval abdomen, so I wrapped two shaped

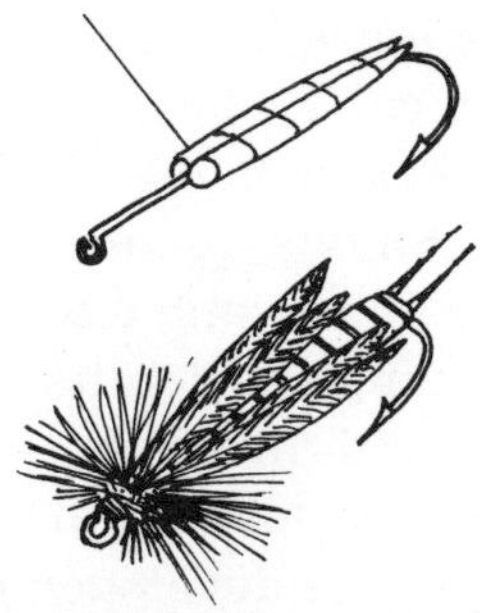

strips of balsa on either side of the shaft of a No. 10 2X long hook, covering them with brown floss ribbed carefully with gold thread to simulate segmentation. I even went so far as to count the natural's segments and dotted the top of the abdomen with dark acrylic paint to further match my model. Stone flies have two short, stout tails that are well separated, and these I copied by tying coarse feather fibers to the outside of the balsa outriggers before the floss and ribbing went on. Next came four gray-black hackle-tip wings tied flat and splaying to the sides, and finally, brassy dun hackle, trimmed on the bottom.

It was a thing of beauty if I do say so, and my stone fly might have won a ribbon at the State Fair, but not if the trout had been doing the judging. They barely gave my effort an honorable mention; in fact, it was more of a fishy Bronx cheer. I had my youngest son, Jimmy, along that day, and to rub it in, they were taking the little fellow's crude nonimitation pretty well. I'd taught Jim to tie a kind of primitive Royal Coachman with a flat hair wing (called a Betty McNall locally), and by splashing the fly around, he was catching about two to my one. And when I further trimmed the hackle on my Stone Fly to just a few sparse strands and imitated his violent presentation technique, they couldn't have cared less. I can remember feeling more than just disappointed; I was a little angry with the Guadalupe's browns. How would you feel if you had spent most of an hour on each artificial only to be left with half a dozen impotent if elegant imitations? Of course, there was no point in getting mad at the trout, since it was I who had tied the sexless Stone Fly. Still, they clearly preferred Jim's ragged brown-white-and-red creation, and I saw no reason to continue the torture.

Incidentally, it's obviously a good idea to fish with one or several companions when you are pattern-testing, and it's probably best if they are *less* experienced or expert than yourself. You see, if it's the other way around and if the other person is having a better day, there's no way to know that he (or she) isn't simply reading the water more perceptively or presenting the fly in a better way for the conditions at hand, etc. Alternatively, if the more junior angler is doing less well with standard patterns, you can trade and let him try your experimental fly. If this helps him, you know you've possibly got something.

Anyway, since the "correct" fly hadn't done the job for me, I was in the mood to try something ridiculous that afternoon, and, probing the depths of a box I carry for especially far-out flies, my eye caught a flash of pink. It turned out to be the wings of an unlikely looking artificial that would have otherwise been a Grey Wulff (as developed by Lee Wulff), a nameless and forgotten relic from former days of fanciful flirtations with pastel wing colors.

The first float was good for a 13-incher, and my log shows that the next six pieces of target water yielded four fish. I knew it wasn't a fluke, either, because they were going out of their way to take the flamboyant fly coming across current and boiling up from the depths of runoff-roiled pools to engulf it. But now I had a very clear question that seemed important to try to answer: How would a kosher Grey Wulff perform in comparison?

The pink-winged version seemed only slightly superior until about four p.m., when a bewildering complex of shadows and spots of bright glare covered the Guadalupe, making fly-watching a near impossibility. From then on, the drab brown-winged Grey Wulff fell behind just on the basis of getting lost from view. Meanwhile, Jim's attention span had run out and he was engrossed in a biological project involving the capture of water boatmen in a tin can, so we never completed the second half of the classical pattern-switch experiment.

I regarded this experience as mildly interesting at the time, wondering if a pattern with pink plumage could have any real potential. I had no particular plan for the fly, but, as so often happens when some new idea is bounding about the recesses of one's mind, it wasn't long before another opportunity presented itself, this time on the Brazos River. Anglers frequently do a double take when New Mexico's Brazos comes up in a conversation, since the Brazos is generally thought of as being "deep in the heart of Texas." Fortunately, there are two rivers with this name, and while Texas has the longer one (what else?), we New Mexicans would never trade, for ours is a marvelous trout stream, good-sized, swift, and clear.

On this occasion, we were fishing several miles below the towering vertical cliffs of the spectacular Brazos Box Canyon, and on three consecutive days were blessed with a noontime spate of a dry-fly action as inspiring as the scenery. For this we were indebted to a hatch of basically grayish mayflies (*Epeorus longimanus,* I believe), and although the insects weren't especially numerous, they clearly brought out the trencherman in the trout. All was quiet until late morning, and then, as the first of the duns struggled on the surface, the whole river would come alive in less than fifteen minutes as browns and rainbows alike took our floating artificials with zeal for a period of nearly two hours.

It wasn't the kind of situation in which you riffle frantically through fly books in search of something that will work—quite the contrary. As I recall, an Adams got the first shot, followed by a Golden Adams, a Laramie, a Grey Wulff, and finally an unweighted No. 14 Maniac Nymph presented just deep to the surface. They all performed with distinction, so there was no need to gamble with an unproven pattern, but the Grey Wulff reminded me of the past month's surprising events on the Guadalupe, and I had tied up half a dozen of the pink-winged version in sizes 14 and 16 for future reference.

Thus, on the second day I tried one, expecting to see a high strike:hook ratio. I predicted a high ratio because the trout were getting a pretty good look through clear water this time, and the fly in no way resembled the naturals apart from size. Therefore, I was surprised a second time when during a thirty-minute period the queerly winged pattern elicited fifteen strikes, hooking thirteen of the risers and solidly enough that twelve trout were landed and released. And when I backtracked to the previous patterns, I was unable to find one that did as well, either that day or the next! Strange but true, once again the Grey Wulff tied with pink wings had been taken with real sincerity.

Well, two experiences do not a pattern make, that's for sure, and particularly since both took place in a state that's not generally regarded as "trouty." I've mentioned a good many New Mexico streams in describing various flies, and it's a good bet the reader has wondered whether our Southwestern clime is really a testing ground useful to the rest of the flyfishing world. I know I have. Not being at all famous, I don't get invited to fish in New Zealand, Argentina, Yugoslavia, or any of those good places. I was delighted to be asked to say a few words about my hobby in Sheridan, Wyoming, recently and as my wife is prone to point out, I'm not a professional angler anyway, so I need to stay around home and earn a living. Beyond this, New Mexicans aren't always sure they are part of the United States in the first place. For instance, a colleague who belongs to a well-known American scientific society was surprised recently to find his name listed among the foreign members! We haven't a single Big Hole or Ausable to brag about, and our one blue-ribbon trout water, the San Juan, needed to be chilled down via a dam to create a fishery, so I have a sort of geographical inferiority complex. It's not all bad, though, for this feeling has created a correspondingly strong urge to try my patterns in more prestigious waters to the north, and I've been pretty compulsive about it too.

Here's what I've found: If a certain fly is less than impressive locally, it's quite unlikely that it will make a worthwhile contribution elsewhere. Alternatively, a pattern that's really good in the land of the chili pepper is almost certain to be productive in the far-off and fabled places. It's a pleasingly straightforward rule and one that applies even to patterns as peculiar as the pink-winged Grey Wulff. Accordingly, the fly has come through repeatedly for me in Colorado, Wyoming, Idaho, and Montana, with relatively few bad outings. Or to put it differently, I've not seen a shock reaction on the part of the trout to the shocking pink wings despite the fact that my boys refer to the pattern as "Dad's gay fly." (I've noticed that they don't scruple to fish with it, though.)

I didn't particularly like the boys' name for the fly, nor was I sure that Mr. Wulff would want his name applied to an altered pattern, especially one with pink wings, so I settled on Pink Wolf as a compromise, since I felt the name should acknowledge a significant debt to his Grey Wulff.

My goal in writing this book included the hope that at least some readers would tie and try a few of my patterns. However, in the case of the Pink Wolf, I'm afraid prospects are dim. It's only reasonable to be skeptical of such an outlandish set of wings, and besides, it's a fair amount of trouble to collect the materials and dye up a whole calf tail knowing that it may only be used to tie a few flies. On this account I felt I should mention some alternative wing materials that don't commit one to doing an entire tail in pink.

A friend who's in the tying business once gave me a patch of white Dall sheep hair. It's nifty stuff. Longer-fibered and coarser than deer, it's also softer and not too fragile. Best of all, Dall dyes beautifully to any color and gives you a very light and full-appearing set of wings (if a bit "spiky"). I was having fun exploring various tying techniques with this hairy gift until I happened onto an account of a Yukon sheep hunt in a magazine at the barber's one day. You know the format; it amounts to an ambush, sniper-style. The magnificent rack appears on the skyline, the scope's cross hairs lock in, the hunter's finger squeezes, there's the reassuring thud of the recoil, and a jacketed slug is on its deadly mission. Inevitably, the sheep catapults into the air and cartwheels down a tundra slope, crimson staining his shining white coat just behind the shoulder (*always* the shoulder; I guess they don't talk about rump hits). Anyhow, I decided I didn't want to be even an indirect part of further executions in the Yukon, so I haven't used Dall since, although this is probably ridiculous—calves have nice brown eyes, too, and their tails don't engender any guilt feelings whatever.

There are many other types of hair that might be suitable for wings, and I think you should try them. For example, elk has become quite popular, although it's too stiff for my liking. Others use tender antelope fibers, etc. (You will find that some of these are difficult to dye unless you start with nearly pure white hair of a kind that will accept the pigment.)

This brings us to a final option, and that is polypropylene yarn or sheeting or other related synthetic materials. Now, many tyers practically gag at the idea of replacing natural feature or hair wings with something man-made, and I generally agree. On the other hand, polypro yarn is inexpensive and comes in practically any color. It won't and can't absorb water aside from droplets that wedge between individual fibers and are held there by surface tension, just as by the bristles of a brush. Of course, the droplets come right out when the fly is blotted and are discouraged from re-forming by silicone flotants, so polypro wings are ideal for dry flies. But what I like best is that polypro is just unbelievably easy to tie with. There's no doubt that many readers will be more skillful than I, but how many can get a good hair wing on a No. 18 hook? And how many can predictably turn out a perfect pair of hair wings on a No. 16? The 16s I can handle on a good day; that's my limit, though, and even then I can seldom stand to tie more than three of them at a sitting. I have a reputation for extreme irritability at such times, too; even our dog knows this. But when I'm working with polypro, things are better. Polypro ties in with almost no lump at the wing base. In the case of hair, there's a damnable bump or hump that gives the best tyers pause and the rest of us fits! This is because the lumpy wing butt needs to be covered up by the hackle, and yet it forces the fibers to splay every which way unless you're both careful and crafty. Also, I've never had polypro wings spin or twist on the hook shaft, a real threat with less than well-tied hair wings, and a calamity that results in terrible fly balance plus a lop-sided float. The synthetic wings don't even need to be matched for weight or length, right versus left, since they are trimmed for length and thickness after they are tied in. Because the yarn flares away from the point of attachment in the shape of an inverted cone, it's no trick to shape and taper the ends as you wish, thinning the tips to achieve relative translucency. It's just neat stuff to work with.

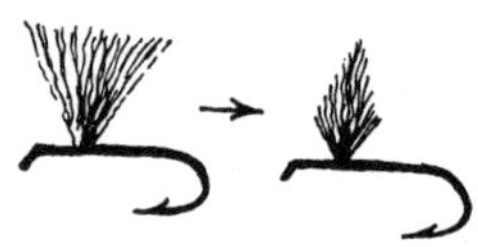

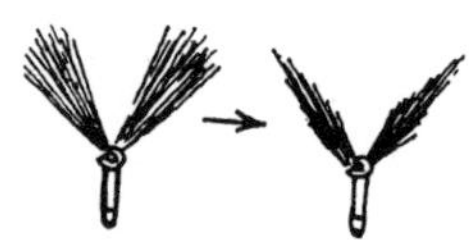

The above remarks apply to any wing color in polypro and to any pattern, but despite the favorable comments, I still tie many 16s with hair wings and very seldom use polypro on No. 14 flies and never on larger ones. The reason is that, individually, the fibers in the synthetic material are far finer than hair of any kind, and with the thicker, heavier wings that go with larger hooks, you tend to get a matting effect; each wing has a solid, opaque appearance that's not all realistic. Also, in sizes 14 and above the polypro wings don't sit up on top of the surface film as well as hair wings, indenting it like the bottom of a boat, so there's little flotation advantage.

POLYPRO *HAIR*

I've gone on at some length about the Wolf's odd wings

without so much as mentioning the pattern's hackle, body, or tail. I use my standard gray/brown deer-hair tail and ordinary blue-gray fur dubbing for the body (dark muskrat) and thus have nothing very special to say. As far as the hackle is concerned, I prefer medium-blue dun or a shade that's just to the dark side of medium. I've also picked up an idea from Jack Dennis' fine book *Western Trout Fly Tying Manual* that I like very much. He starts out with badger hackle (whitish edges and dark center) and dyes it to a blue dun. The result is a hackle that's lighter at the edges and quite dark in by the hook and wing roots, just as an insect's thorax should be most opaque centrally. (There's more to come in Chapter 11 on photodyeing materials to get a dun color.)

Frankly, I find it difficult to conclude this chapter on other than a "so what" note. As I said several paragraphs ago, it's doubtful that there will be many takers when it comes to actually testing such a garish fly. However, my motive was to try to show that a bizarre wing color won't necessarily ruin an otherwise sound pattern, and at times, it may actually enhance its effectiveness. Thus, my intention was simply to encourage bold (or foolish) experimentation. "Think pink," for example, and conjure up a vision of a Royal Coachman tied with blushing wings; what influence might they have, pro or con? And getting back to Lee Wulff, I'm not so sure he wouldn't approve. It seems to me that Mr. Wulff has done more than anyone else to introduce fly patterns that help to get anglers out of the traditional construction rut. He has long championed the idea of showing the trout something startling in the way of color or configuration. The blocky "mouthful" Wulff flies are typical in this regard, and their continuing popularity proves his point, I think.

CHAPTER 10

The Blue Max

The Max is actually a very old member of my fly family and yet, in its present form, one of the newest. In all honesty, I believe it's an exciting pattern and one that has unusual potential. The Max's heritage goes back to the fascination I have for mixing hackle colors. If you check over color plates illustrating dry-fly patterns, it becomes apparent that the majority employ one of three general hackles: brown (dark to ginger), blue dun in some shade, or grizzly. You could argue that there is a big difference between chocolate brown and ginger or between iron-blue dun and pale dun, but still, the basic hue is the same. When I categorized sixty-five consecutive flies from the plates in Ray Bergman's *Trout* in this way, I came up with 40 percent brown-hackled patterns, 20 percent blue dun hackle, and just over 15 percent grizzly hackle. Thus, about three out of four flies fit the formula, while the others sported black, white, olive, yellow, or badger hackle, but with few patterns in each group.

As I've explained, I became convinced early on that the grizzly/brown hackle blend of the Adams was somehow magically titillating to trout, and I became convinced that other mixtures of the three commonest colors should be tested. This eventually led to the Laramie Grey's blue dun/grizzly hackle and to a whole batch of patterns with a marriage of brown and blue dun feathers, the "B&B" series as I called them. And the idea really made pretty good sense since "grayish brown" is a description that could be applied to all sorts of insects whether they might be mayflies, caddis flies, stone flies, grasshoppers, or whatever. Unfortunately, the idea was more appealing than the results, and in the interest of saving lots of words, I can sum up about twenty years' worth of experimentation by admitting that my B&B patterns were none too good. They weren't really that bad either; most of them caught fish all right, but as we've agreed before, if a pattern can't make a substantial contribution to your armamentarium of artificials, why bother?

The reason I kept at it year after year was that the B&Bs looked so pretty in the vise and on the water, too. I'd sometimes tie three or four models in a single season, winged and wingless, tailed and tailless, and with various bodies of floss, tinsel, quill, dubbing, herl, clipped deer hair, and so on. Although I played around with form, shade, texture, surface sheen, etc. for nearly two decades, there continued to be a chronic gap between my expectations and the performance of these fetching flies. From time to time I'd come up with such a promising pattern that I'd tie a whole batch before any sort of trials were conducted, and at one point I had several hundred B&Bs roosting in an old cigar box, a futile flock of fur and feathers. I eventually concluded, correctly I believe, that merely mixing brown with dun hackle doesn't automatically result in a winner. Apparently, if I was going to come up with something new and worthwhile, the B&B hackle would require positive integrated input from the body and/or wings in order to reach a "critical mass" from the standpoint of impressionism.

In the end, the Max sprouted blue wings (of all things), and I'm not particularly proud of the way it happened. Over the years, I'd tried and given up on both hackle tip and hair wings in blue dun and also in plain gray from the palest shades to the darkest. However, I'd never thought about putting really blue wings on a B&B chassis; I felt the place for bright blue was in brookie patterns or on salmon or saltwater flies. I'd like to tell you that the genesis of this pattern's wings was the result of insight and logic on my part, but the truth is that the catalytic spark was a pure accident. I was sitting at my tying bench in the study with a blue dun and a red game neck in front of me one day, waiting for an idea, when my daughter Elizabeth dashed in seeking protection from one of her older brothers (a common occurrence). In the course of dispatching her pursuer, I noted that the end of each pigtail was secured by a strand of sky-blue yarn. At least this was something I hadn't tried before, so detaining Elizabeth briefly, I snipped off a sample. The result was not particularly attractive, for the bright wings looked out of place against the relatively drab hackle, a dubbed body of muskrat fur, and deer-hair tail. In fact, this time I tied only one fly, a No. 14, and tried it occasionally during the remainder of the season. This bright-winged B&B performed well enough, if not spectacularly, so the next year I tied nine copies, three each in sizes 14, 16, and 18, using dyed calf tail or polypro yarn. I didn't give this cerulean creation a very high priority rating in terms of stream trials, although as data accumulated, I began to suspect that this version of the B&B might be different, and in a most desirable way. Specifically, it seemed to come to the fore best on days when the trout were suspicious and skittish so that strike:hook ratios with other patterns tended to be high. I realized that if this was a true attribute, I'd really have accomplished something, for solving unfavorable ratio situations can mean the difference between a good and a bad day.

The way things worked out, the fly didn't stay as it was very long, for it soon developed an unusual hedgehog body. I suppose my Scottish heritage was largely responsible. In a moment of self-imposed weakness, I convinced myself that I had dire need for a natural blue dun neck, and, of course, this cost me in excess of $40, almost six times more than I had ever paid. It was a fine piece of fowl, too; I relished every feather; until, as happens to all necks, it eventually wore out. As a general rule, I do most of my dry-fly fishing with No. 14, 16, and 18 artificials, so my denuded necks still have a plethora of No. 12 and larger hackles remaining. I tend to have trouble throwing certain things away, and the process of bagging that exhausted blue dun neck with moth crystals was peculiarly painful. There just ought to be some use for these springy, glossy hackles, I thought. And suddenly a penurious plan was upon me: I'd use them for bodies!

I simply cut the fibers from both sides of the tip of a No. 12 feather and tied it in on top of the tail with the butt protruding to the rear, as shown. (The trimming was to get a stubble that wouldn't pull free, just as most tiers trim the butt end of hackle feathers before tying in.) Aside from the tail materials, the shaft was still bare, and I then proceeded to cover it with the working thread followed by close-spaced spirals of hackle, stopping just past mid-shaft, where I tied off and amputated what was left of the hackle feather. This was just like palmer-hackling a body, with two differences: There was no body underneath (just the tail tie-in), and the hackle spirals were as closely spaced as possible. Of course, the thing looked a mess, until I trimmed it to shape, much as one does the body hair in the Irresistible. The body-feather stubble was packed closely enough to give a fairly solid burr-edged silhouette. Since I was mixing brown and dun feathers in the hackle and since I also had huge quantities of larger brown hackle lying around going to waste, I decided to do the body with two feathers, one blue dun and one red game. This produced a color that was mottled and naturally very much like that of the hackle, so that the two pieces of the fly's anatomy came into harmony. Putting the hackle to good use pleased my Scottish instincts, and I also found this body construction to be quite durable. There was no need for a protective overwrap or rib, and better yet, it had excellent floating qualities. The stubby hackle burr trapped air bubbles, especially after dressing with silicone, and was intrinsically light, certainly lighter than the fur dubbing I had been

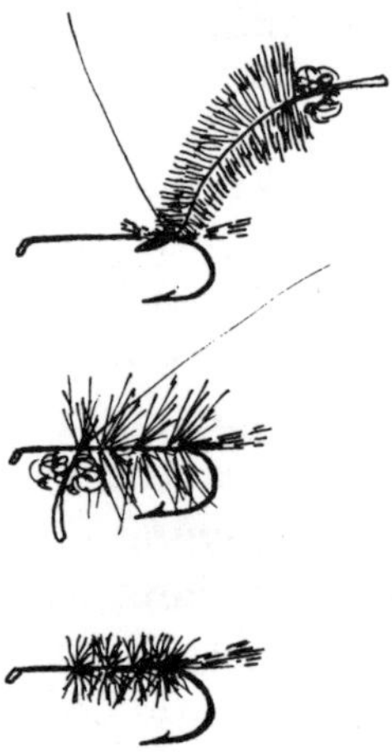

TRIMMED

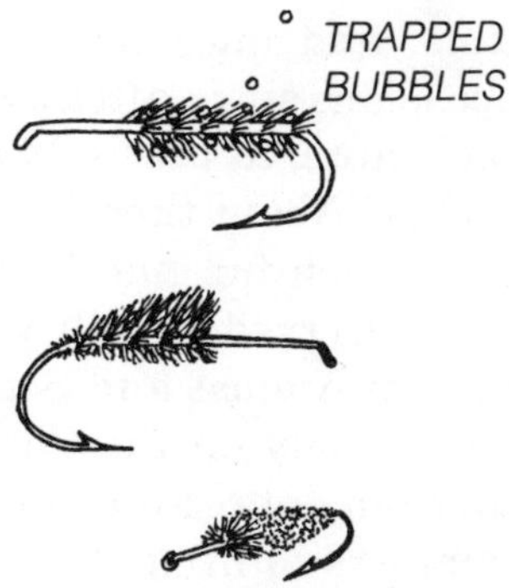

using. Further, I could do all sorts of things with body shape. On smaller hooks I trimmed the underside close and flat so as to leave as much space as possible between the body and hook point. As noted, this is helpful in setting a hook. And if I wanted a sharply tapered body, back to front, or an oval body, there was nothing to it. I'd not seen this kind of body described before Terry Hellekson's enormous collection of pattern recipes was published in 1977. In *Popular Fly Patterns,* Mr. Hellekson lists both a nymph and a dry fly, the Orange Sedge, that employ this clipped-hackle technique.

So this is the story of the pattern's burr body. In any event, the blue-winged fly with its bristly fuselage really seemed to take off, and one evening after a particularly good day on the stream, I asked the family to help choose a name for the fly, presumably something with "blue" in it. It's surprising (although not especially interesting) how many "blue" name combinations you can think of. We came up with three finalists plus a consensus worst name, Blue Monday. I favored Blue Angel, while my wife preferred Blue Beard for some reason. The kids were unanimous in choosing Blue Max, however, on the basis of the popular movie of the name, so democracy prevailed.

The first hint that the Blue Max might be something special came even before the fly got its name on a trip to the Conejos River in southern Colorado. It was the weekend after Labor Day, a part of the season most anglers associate with dry-fly fishing. I think this is also a good time to test wet flies, because a background of low clear water gives the trout an opportunity to render a really critical appraisal. On this day I had planned to look specifically at three patterns, a nymph and two streamers. Unfortunately, the classic Conejos is dogged by a road all the way up to Platoro Reservoir, where this sizable river takes its origin. Still, in a few areas the stream canyons a bit, becoming a little difficult to get at, and I began the day in such an area.

Now, as is no doubt obvious, part of the amusement in fishing for me involves conducting competitive matches between patterns, but it's also gratifying to catch some trout from time to time, and once in a while a conflict develops. It did on this occasion, since after three hours of honest effort with the wets, I had captured precisely six rainbow measuring all the way up to 8 inches! About then it dawned on me that I really wasn't having that much fun and hadn't learned a single thing beyond a little well-deserved humility. It was surely time for a change. I knew that in August and September the Adams is an especially excellent dry fly on the Conejos, customarily finishing first in these tournaments, and sure enough, an Adams soon added some numbers to my catch record. Even then, strikes weren't numerous and the rises were diffident and insincere, so it was still slow going.

After an hour I decided to see if a Blue Max might jar their reluctance, and when the blue-winged fly exactly doubled the Adams' yield during the next hour, I had an unplanned experiment cooking. In this kind of study it's essential to go back and retest pattern #1, since conditions may have changed, and this is what I suspected had happened. For me the Adams sets a standard of excellence; it's a "Hall of Fame" pattern, a formidable adversary at any time and a competitor that had never lost a late-season game on the Conejos. As expected, second time around the champ did better, but when the Max got its second shot, the new fly again won handily. The scoreboard looked like this:

Time	Pattern	Catch
2:45 to 3:45	Adams	4
3:50 to 4:50	Max	8
4:55 to 5:55	Adams	6
6:00 to 7:00	Max	11

A score of nineteen to ten isn't exactly a rout, nor does it really prove anything. It's important not to overinterpret results like this, as I've explained in the concluding chapter. Regardless, this time the figures were just the reverse of what I'd have predicted.

While it hasn't happened often, on occasion the Blue Max has been about the only thing that would produce for me. One of my favorite rivers is the Crystal in the Elk Mountains of west-central Colorado. On two occasions during separate seasons I would have been practically shut out on the Crystal were it not for the mysterious lure of the Max. If the rawboned lore of the Wild West appeals to you, the valley of the Crystal is hard to beat. Guarded by lofty Mount Sopris at its mouth, the 14,000-foot peaks at the river's headwaters are pocked by fabulous mines. These timberline tunnels gave up fortunes in silver, gold, and lead, and here there is also a veritable mountain of marble that fed what was once the world's largest marble quarry and mill complex. Much of the drama is related to the terrible blizzards, avalanches, floods, mudslides, and fires which seemed continually to beset the resourceful miners from the district's beginnings a hundred years ago. Despite these perils, immense slabs of marble for the Lincoln Memorial, the Tomb of the Unknown Soldier, and many other famous edifices were carted out of the valley via a spur line off the Denver & Rio Grande Western. The train followed the Crystal's banks closely all the way down. The right-of-way is barren now, its tracks scrapped during the war effort in the '40s, but multiton chunks of statuary marble remain in the railbed, placed there as ballast, while others protrude from the water's edge where they were dumped to help hold the river to its course. It's strange to look up at these glistening, sharp-sided forms as one wades along, for their pale, angular austerity somehow bears witness to a romantic bygone era.

It's not that this has anything to do with my angling difficulties on the pretty Crystal; quite the contrary. As this historic valley has progressively opened its doors to tourism, the river has become something of a put-and-take proposition with sizable stocks of rainbow, but for some reason I still have problems. In search of larger trout, we usually work the heavier water downstream from the startling crimson cliffs that surround the old coal camp of Redstone, and it was here that the Blue Max came through twice after my other old reliables had faltered. On neither occasion was there appreciable surface-feeding activity, so I wasn't matching any sort of hatch, and since the fish weren't exceptional (save for one gorgeous cutthroat), I won't go into detail.

You might theorize that the Max was successful on the hard-fished Crystal because it was unfamiliar to the denizens of the river. I doubt that this was the whole story, though, since the fly has pulled essentially the same trick under totally different circumstances where the trout were inexperienced and practically virginal as regards anglers' solicitations.

Way up high, Wyoming's Green River does a 180° turnabout some miles below the big lakes at its head and then continues on southward along the west face of the Wind River Range. As the bend begins, the Green picks up a major although seldom visited tributary known as the Roaring Fork. Getting there involves a lengthy trek over the faintest of four-wheel-drive trails atop a high bluff overlooking the river. It's a big arresting vista, full of soaring hawks and curious antelope with the bald, gothic summits of the Wind River Mountains off to the south. The driver can ill afford to sightsee, though, lest he smash an axle on the dome of a grass-shielded boulder, and it's slow going until the stream finally appears ahead, cascading down a series of chutes and cataracts from an open bowl of high peaks. One more steep pitch and you emerge into the most incredibly beautiful meadow I believe I've ever seen, bright with wild flowers and flanked by belts of dense timber. A backdrop of jagged granite sets off the broad expanse of green lawn like a jewel, and through its heart meanders a serpentine trout stream. The Roaring Fork is about 20 feet in width and replete with lush undercut banks along its convexities where the water is deep and green with beds of water weed. It's strictly calendar art and an obvious setting for a fly-fishing orgy.

The first time I saw the place it was evident that the meadow had not hosted many eager anglers, for upright grass in the "road" tracks and a faded antifreeze container suggested that the last guests had been hunters the previous fall. As we rigged up, the question was not if, but rather how many, how large, and which kind, if you see what I mean. This sort of preconception explains why the three of us stared at one another in numb silence some two hours later, for no one had so much as seen a fish. I'd offered them Muddler Minnows and white marabou streamers while one companion had bounced a bunch of nymphs along the bottom and the other had plied his dry-fly wares, all to no avail. The Max? Well, nothing special, although an average of eight trout per hour is somewhat better than none, and a few of the brookies were very nice indeed. We didn't happen to regroup until it was time to leave, and sadly enough, both friends barely avoided taking the collar on the day, so I gave the Max considerable credit.

A few closing comments concerning construction details are in order. The wings should be a pale sky blue and not a bright iridescent hue. I suppose the adjectives "powder" or "pastel" would be appropriate for the shade I've found to be most effective. My standard deer-hair tail looks comfortable on the Max; the mottled gray-brown fits right in. As to the body hackle, it makes no difference whether you wrap on the brown or the dun feather first; they get all entwined anyway, and I often wrap them in tandem. I start with the tips of the feathers rather than the butts because the tip is more pliable and easier to wrap. The butt almost always gets trimmed off. You'll find that a No. 10 or

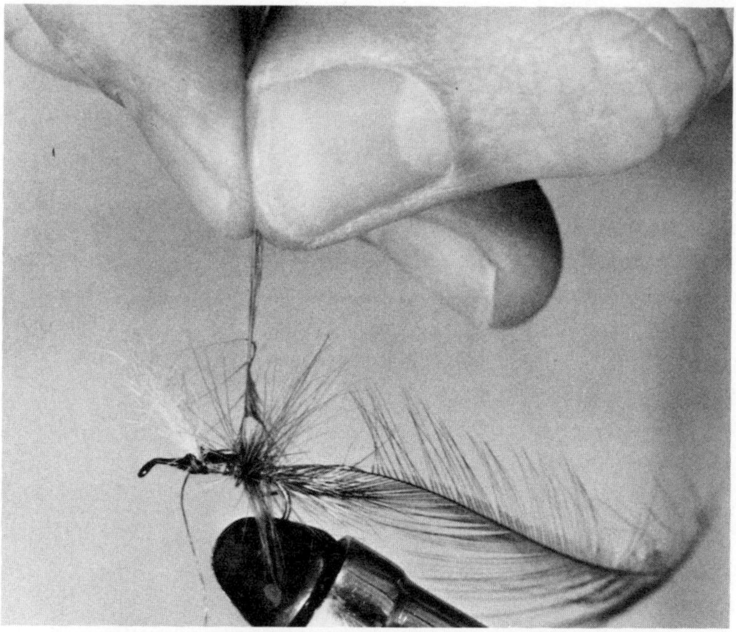

In wrapping on the body hackle for the Blue Max, it doesn't matter whether the brown or dun feather comes first since they will be entwined. Here, the author is tying in body hackle, with the second feather to come.

12 feather is too stiff, even at the tip, for hooks smaller than a 14, and I get around this by saving smaller hackles of inferior quality from a neck for this purpose. Even the best necks have a few badly shaped or somewhat webby hackles that are perfect for Max bodies. Hackle color balance seems to be quite important, and here there are two considerations. In the first place, I'd definitely stay away from dark browns and duns (iron blue); a medium or medium-to-light shade is far preferable. Second, just as grizzly dominates the blue dun component in the Laramie Grey, so does the brown dominate the dun hackle in the Blue Max. I've learned to shoot for a 2:1 or 3:1 mix, favoring the dun member of this impressionistic blend.

The conclusion to this chapter ought to include a lucid and believable explanation as to what makes the Blue Max go, and this is an embarrassment for me. On the basis of a whole series of observations, I've concluded that I don't have the slightest idea how the trout perceive the Max. I really doubt they've given it much thought. For instance, you might suggest that the fly's prickly outline makes it a clever caddis copy. Perhaps it is, but I've never been able to make a correlation between what the Max accomplishes or doesn't accomplish and the presence or absence of caddis. It's hard to imagine that the wings give the Max unusual visibility from the trout's standpoint. At least for the angler the powder-blue wings can't begin to compare with white or gold in this regard, and the rest of the fly's anatomy usually fades smoothly into the watery background. Although I don't want to get into the ancient argument about whether trout appreciate color and if so, how, something about the blue wings seems to matter. I've wondered if the color somehow blends with the sky, as seen by the trout from beneath the surface, in a way that's special in terms of realism. But if this is so, the Blue Max should be more of a hatch matcher than an attractor or impressionistic pattern, and yet most of the success I've had with the fly has been under hatchless conditions. Thus, I remain in a blissful state of enthusiastic ignorance!

CHAPTER 11

The Tadpole

I'm a staunch supporter and advocate of the streamer; it's a fly for big fish. Like most anglers I have a fantasy to the effect that somewhere, someday, I'm going to catch a perfectly enormous trout, and I usually work a streamer into the lead role in this sort of daydream (and at night, too). The dreamed sports-page headline reads: "Record Trout for Local Fly Fisher." Just between us, I usually return the giant fish too, since there always happens to be a group of highly reliable witnesses watching from the bank, one of whom invariably has a tape measure and precision scales along; if you're going to fantasize, might as well go first class!

I suppose there is some of the Walter Mitty in all of us, and as a result I've copied, altered, or otherwise tampered with innumerable streamer and bucktail patterns during the past thirty years in hopes of developing a fly that would be omnipotent in the manner of the "perfect nymph." However, I gave up on the project some time ago, having convinced myself that the universal streamer is even more of an impossibility. The basic reason is that trout seem to respond to streamers in a way that's highly variable and difficult to predict. There are undoubtedly a whole bunch of reasons for this, some of them clearly beyond our comprehension. Take the Muddler Minnow. The Muddler must come about as close to the ideal minnow copy as any pattern can, and I've tried it under virtually all conditions and on almost every piece of water I've fished during the past fifteen years. With all due respect, and a lot *is* due, the Muddler's performance has been all over the grade scale from A to F.

THE MUDDLER

Part of the explanation is geographical and probably relates to the sculpin minnow, that little fish with the big "stepped-on" head. In the Jackson Hole/Yellowstone/Montana country where sculpin abound, the Muddler is an absolute must. Yet I'm told that there are sculpin in many trout streams in Wyoming, Colorado, and New Mexico where I'd rate the Muddler as merely a very solid, consistent streamer lacking deity status. Meanwhile, in certain West Coast waters some have found this best-known of all the modern streamers to be a disappointment. And even within the heart of Muddler country up north, you'll see very severe alterations of the original pattern. These regional favorites tied with white or yellow marabou wings in place of the drab turkey feather hardly resemble the parental Muddler.

You could plow differences in trout species into the equation by suggesting that the rainbow, cutthroat, and always amiable brookie are better streamer customers than is the browns. In a very general way this would fit my own experience, although brownies are notorious cannibals and it's interesting that Jack Dennis, a Jackson Hole professional, states in his *Western Trout Fly Tying Manual* that browns are particularly attracted to the standard Muddler. So there's more to it than this.

Getting away from the Sculpin specifically and referring to just any small fish as a "minnow," I used to think that one ought to be able to compute the likelihood of success with streamers by noting the relative abundance of little fish in the shoals along banks. It's a logical enough thought, but I've gotten no mileage out of it whatever. I can think of a set of five streams, two in Idaho and three in Colorado, which are cold, swift, and apparently about as poor in minnows as trout streams can be (they obviously must contain tiny trout) but where streamers are just excellent. For that matter, it isn't entirely rare to hook an exhibitionist with a 5-inch minnow sticking out of its maw, tail first, like a giant cigar; you wonder how there could have been room for the fly. The other side of the coin brings memories of more tepid, languid waters teeming with succulent sardines in which I do ever so much better with dry flies and where I've yet to find a partially digested minnow in a trout's craw.

So the trouble is you've practically got to know the stream in question through experience or by reputation, and even then seasonal or weather changes influencing water level, current strength, and clarity can scramble things. For instance, a cloudburst can turn a strictly dry-fly setting into an excellent streamer situation in a matter of minutes. To complicate matters further, the kind of presentation you choose is most certainly an important factor as regards trout reaction. Just think how many different ways a streamer can be fished. The numerous options begin with whether you want to work the slow water or the currents, and if you prefer the latter, whether it's better to pull the fly across the current or directly back against it. Then the retrieve may be fast, slow, or in between, and smooth, jerky, or in between, and there's still the matter of a shallow versus deep presentation to consider. Or you might just wish to show the artificial floating naturally, dead-drift, as an exhausted minnow might present, off an upstream cast. The traditional approach with streamers is the cross-current swing, although based on my own variable experiences; I hate to recommend any single technique to the class I teach in flyfishing at our university. The point is that it may well be the angler who's at fault when his streamer fails to impress rather than the pattern.

Having gone to some length to convince you that designing an all-encompassing minnow copy is an unrealistic goal, I've also managed to paint myself into a pretty tight corner, for the Tadpole clearly belongs to this family of artificials. I have an escape plan, though. I hope to defend the Tadpole on the basis of the following arguments:

1. That statement in the opening paragraph about streamers and big fish is perfectly true, and consequently, these flies have considerable headhunting appeal. Admitting that "Superstreamer" is a myth, when trophies are at stake even a pretty good facsimile should be a welcome addition to one's flybox, and the Tadpole represents my best effort of this type by far.
2. The Tadpole gets its color in a unique way that I hope you'll find interesting and potentially useful.
3. Although the pattern's other individual construction characteristics are not so unusual, taken together they nonetheless provide considerable opportunity for further experimentation.
4. The Tadpole very probably suggests food forms other than minnows, including nymphs (and even tadpoles). Thus, I suspect the fly has a wider imitative scope than most streamers.

Plagiary is not generally regarded as an altogether desirable activity. However, when practiced with scrupulous selectivity, carefully disguised and carried out with tact, I suspect that it can pay off. There are temptations for the fly tyer in this area. Now, assuming a person has committed such an offense, does an open confession lessen the crime? I feel better about things, anyway, so let me freely admit right now that the Tadpole is "indebted" to three well-known streamer patterns: the Muddler Minnow, the Matuka, and any of a number of flies tied with a marabou feather wing.

The story begins with the Matuka. I have always been bothered by the tendency of the wing(s) on standard streamers to part company with the body as the fly is seen side on in the water. Because a streamer's wing is intentionally light and the body is necessarily much heavier, the plane of the two will separate except in a good swift current or during a brisk retrieve when the wing gets plastered back against the body. Otherwise the body will naturally hang below the plane of the wing in the

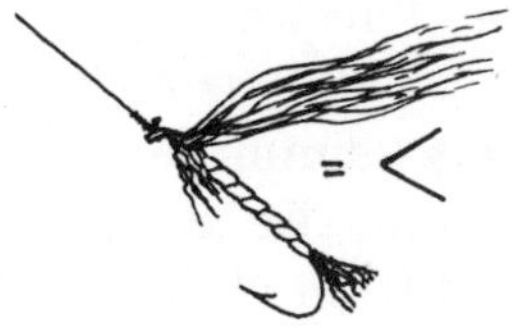

manner of a tipped-over letter V. This peculiar appearance is accentuated if the body has been slightly weighted. Now, a good streamer wing is supple by definition; it ought to move about freely in the water, as the name "streamer" implies. However, the rigid body cannot very well share in the shimmy, and this is why the "V on its side" presentation looks so unnatural. I feel that as far as possible, a streamer's several parts ought to move in sync, just as a real minnow's do. This is a problem to some degree for almost all bucktail and streamer patterns, since the wing is attached only at the fly's head. This is why I was quite interested when I first saw a Matuka several years ago.

Characteristically, the Matuka's wing is stitched down to the top of its body by ribbing, thus causing the fly to become a single unit. The top-mounted wing is still free at its rear tip, so that part of the feather wiggles in a convincingly tail-like fashion, while

The Tadpole is a modification of the Matuka prototype. Here, it gets its hair cut into an oval. This oval-headed, flat-sided streamer swims like a distressed minnow.

the rest of the wing can still ripple despite being anchored at its base. The analogy to a sailfish's sail is quite obvious. More realistically, from the trout fisherman's standpoint, the wing closely simulates the dorsal fins of certain minnows, for example, the sculpin. As I understand it, the term "Matuka" comes from New Zealand and is generic in that it applies to this special wing construction rather than to a particular color, kind of body, or exact type of wing feather. (A tinsel rib to hold the wing on the body appears to be standard.)

I started experimenting with fur dubbed bodies in various colors with either a badger or grizzly hackle wing and found these Matuka prototypes to be serviceable streamers that certainly caught their share of trout. But by nature I couldn't leave well enough alone for very long and soon added a fat Muddler head of clipped deer hair. There was nothing particularly innovative about this modification, since I gather that many tiers ascribe a good part of the Muddler's winning personality to its blocky head construction, and there are just all kinds of good hair-headed streamers around by now. The Matuka's dorsal wing makes the fly thicker from top to bottom than from side to side, so I decided to maintain this fishy shape by trimming the head into a corresponding oval. To my surprise, I found that this head did more than just change the fly's appearance a little. In concert with the sailfish wing it also caused the Matuka to behave erratically in the water. Held in a current or stripped in briskly, the flat-sided fly would twist about the axis of the leader, and dead-drifted in a current it had a tendency to tumble, all movements that I liked to think a minnow in distress might make.

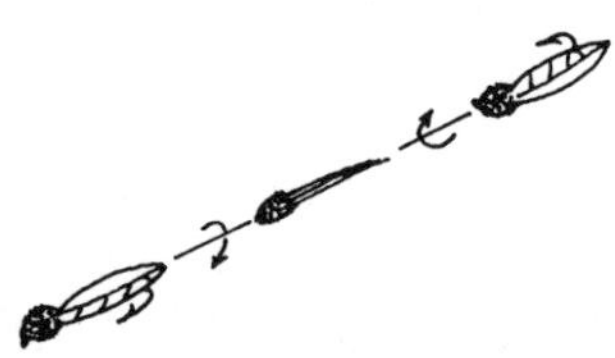

Next came a change in the wing feather. I've always had a

thing about marabou streamer patterns. Nature has no other material in her storehouse, feathers or otherwise, that flutters so seductively in response to the most minor of currents. This is probably a consensus opinion among tyers since marabou has been a popular wing material for a long time. The Matuka's hackle wing had movement, too, but not this kind of sinuous undulation, so I tried the substitution. It wasn't an easy one. As many readers know, individual fibers in a marabou feather are like thistledown, so delicate and soft that they get all caught up in things, and on this account ribbing down the wing was most tedious at first. However, as I gained a little dexterity, it all began to seem worthwhile, for now the fly was beginning to develop some real personality.

MARABOU

The little sketches of Matuka wings I've drawn are out of proportion in that the wing should be longer—in fact, about as long again as the hook shaft. Due to what is probably a silly phobia, I worry that flies like this will get nipped primarily in the rear end where there's no hook, and for this reason have preferred to keep my wings on the short side. The standard Matuka actually needs an extended wing, because hackle feathers are not terribly supple due to the stiff central quill, and a longer, trailing wing gives better action. It's not this way with marabou. Top-grade fluffs have a central quill so thin it's hard to find among all the downy filaments, and even a stubby feather will still do its thing. For that matter, marabou is so soft that you have to choose a slightly stiffer feather than most tiers like in tying the Tadpole. Otherwise, the wing will collapse rather than standing up, fin-like. I clip the fibers from the side of the quill that's going next to the body, as shown, leaving the feather's tip intact as the trailing end.

In the beginning I combined the marabou wing with a gold Mylar body, pressing the Mylar tube flat to conform to the shape of the head and wing. This was a flashy fly, what with the broad sides of polished metal, but it wasn't pleasing. I always felt the

Mylar Matuka came closer to being a spinning lure than a fly, so I went back to natural materials and to a process I call *unit dyeing*.

Assume that you have a favorite streamer or nymph to which you wish to add a particular color cast. I'm not talking about changing the basic color so much as adding a tint, just as ladies (and some gentlemen) will to their hair. Of course, it's necessary to start with a pattern that isn't too dark. The Muddler Minnow is a good example, and let's say that you want to give it a touch of yellow or olive. Logically this would involve predyeing a turkey feather for wings and a swatch of hair for the head and hackle. You'd expect the lighter parts of the mottled feather to accept some pigment, and also the grayish hair fibers. After everything was dry, you'd just go on to tie the tinted Muddler. Unit dyeing is a lot less trouble. I simply take an ordinary Muddler, all finished and ready for the leader, and dump the whole thing, hook and all—the whole shooting match—right in the dye bath.

Now, here's what's special about unit dyeing: It gives the fly a basic unifying color theme with a whole set of subtle variations on that theme. Going back to the Muddler, every part of the pattern will pick up a little dye except the tinsel body. The change in the darker areas of the turkey wing and tail is very slight, that in the head and hair hackle more evident, whereas the lightest parts of the wings and tail will show the new color quite plainly. It seems that Nature colors many creatures this way. Among her trout, the topside and upper flanks of the brookie display all shades of olive, the brown quite reasonably shows us browns with more or less yellow admixed, and the rainbow has steely grays bridging silver to black. Sculpins, shiners, and other minnows are not different in this regard.

I liked the unit-dyed marabou Matuka in olive best of all until I got photodyeing, and then the Tadpole went through its final metamorphosis. Before reading Eric Leiser's *Fly-Tying Materials,* I had the idea that photodyeing must be pretty complicated. I thought that setting up would likely be costly and suspected that the process itself could be tricky in terms of impregnation, exposure, and developing techniques. It's not so. Exposure isn't even part of the process, and the most common chemicals, used by professional and amateur photographers alike, are all one needs. My laboratory does some photography involving the printing of ordinary black-and-white negatives, and I found that used developer and fixer that were ready to be thrown out worked just fine. The only chemical I had to buy separately was silver nitrate.

Here's how I photodye a hair-head Matuka with a Muddler wing to create a Tadpole:

1. Start with a fully tied fly with a pale fur dubbed body and head of light-colored hair, but *do not* apply head cement.
2. Soak the fly thoroughly in tapwater.
3. Immerse it completely in a fresh solution of silver nitrate for 15 minutes with occasional stirring or agitation. (Solution strength is discussed below.)
4. Rinse briefly (several dunks in tapwater).
5. Immerse in developer with occasional stirring for 3 minutes.
6. Rinse briefly in tapwater as before.
7. Immerse in fixative for 10 minutes.
8. Rinse for 30 minutes in running tapwater.
9. Dry completely and apply head cement as usual.

The depth of dun coloration could theoretically be controlled at several points along this sequence. The strength of the silver-nitrate solution, time in the silver nitrate, the strength and temperature of the developer, and time in the developer bath are all possibilities. Actually, only the strength of the silver solution is worth worrying about. Fifteen minutes is enough to saturate materials with the silver salt, and several minutes in developer (same concentration as for developing standard prints) at room temperature gives near-maximum dyeing effect.

I get the shade of gray dun I want by varying the strength of the silver-nitrate solution as follows:

Strength of Silver Nitrate	*End Color*	
1%	Pearl Gray	(pale or light dun)
2%	Medium Gray	(medium dun)
3%	Charcoal	(dark dun; iron blue)
4%	Black	

For the Tadpole I use a 2.5 to 3.0 percent solution to get a darkish dun that's nearly black in the water.

I can't promise precisely how your Tadpoles will come out, because the end result depends on what you start with. An absolutely pure-white material will yield a bluish dun, but if there's just a hint of brown in the material, you'll likely get an olive gray, i.e., a gray with a hint of green/brown. Not that this is bad—indeed, many of my Tadpoles come out this way, because I often use fur dubbing for the body, and sheer-white fur is uncommon, as is pure-white hair for the head. I ordinarly use Dektol developer, and this perhaps gives more of a bronze cast than another common developer, D-76. These minor variations can be important in dyeing a neck for dry-fly hackle, but are hardly noticeable in a streamer or nymph.

You've probably wondered what happened to the Tadpole's rib, for a Matuka-style fly ought to have one. There are several explanations. As I mentioned, marabou is difficult stuff to work with, and I had lots of trouble getting the tinsel rib on straight and tight at the same time. Also, I came to prefer drab dress throughout for the Tadpole, no flash at all. So, on both accounts I substituted heavy (4-0) white tying silk as a rib to hold the wing down. This way, sloppy ribbing spirals accepted the dye along with the body and thus were hidden, while the thinner thread was also far easier to work with than tinsel when cinching down the wing.

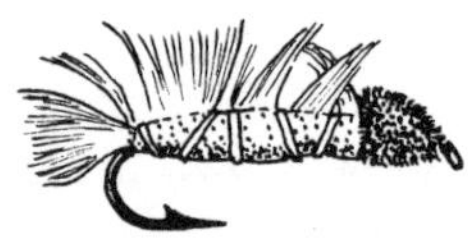

A couple of words of warning about silver nitrate: It's a potent and indelible stain for flies, and it will work every bit as well on fingers, your clothing, the sink, and the floor. Your garage isn't a bad place to work, or better, someone else's garage. Reagent-grade silver nitrate is expensive, about $5 an ounce; however, a little will go a long, long way and several tyers could easily split the cost of a small bottle. (I've also been known to steal a dab from a kid's chemistry set on occasion.) For a 2.5 to 3 percent solution I use one-quarter teaspoon in one-third cup water (2-plus grams in 80 cubic centimeters, more or less) and have found it much cheaper, simpler and less messy to photodye a bunch of flies at one sitting, up to a dozen. This amount of solution will cover the flies easily in a paper cup, and afterward I throw the solution and cup away together with a plastic spoon used for agitation. (Silver nitrate in solution is hard to store successfully.) The developer and fixer can be purchased inexpensively in small quantities from any photographic shop and can be stored, or you may have a hobbyist friend who'll give you some in return for a fly or two.

I think the Tadpole is a much better fly with a little weight added to the hook, for the same reasons as explained in the chapter on the Trashliner series of flies. Further, I like to add more ballast to the front half of the shaft than to the rear. This causes the fly to drift with its nose tilted down in the manner of a foraging minnow. It may be my imagination, but I don't feel that trout react as favorably to nose-up minnow copies. Naturally, when you bring a fly in fairly close, getting ready to lift it out for a fresh cast, the taut leader will tend to make it stand up on its tail even though the path of the fly's travel through the water is basically parallel to the surface. The weight in the Tadpole's nose will tend to keep it down.

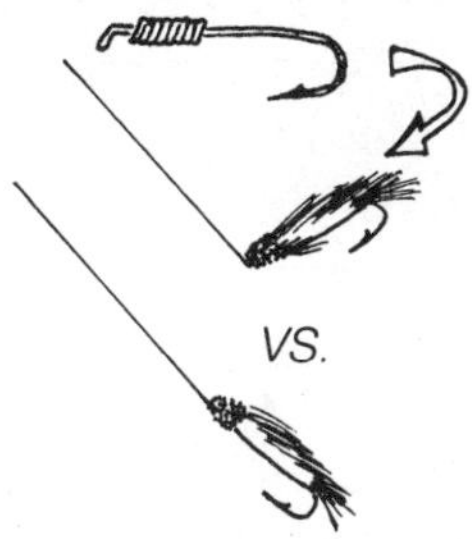

The lead wire also helps to fill out the body, and I like a plump torso for the Tadpole. Conventional streamers tend to have pencil-like bodies. However, they also carry long, trailing hackle or secondary wings that fill out the silhouette, like the clothes on a stick scarecrow. The Tadpole has only the one wing and is scantily clad otherwise, with no tail and a skimpy hackle represented by a few hair fibers from the head which are allowed to go untrimmed, protruding to the rear. As to the body material, I've used grocery string, cotton yarn, and almost every kind of fur scrap. There's no need to buy anything special.

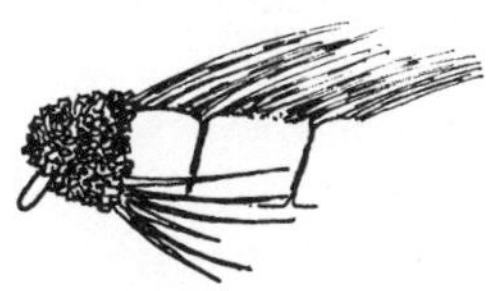

Either a 2X or 3X long hook such as the Mustad 9671 or 9672 is fine, and I prefer them to the really long 6X streamer hooks. Remember not to make the wing too long, either; a somewhat chunky configuration seems to be most effective. Incidentally, an unduly elongate wing will also get twisted under the bend of the hook where it's largely immobile, not a desirable situation.

In the preceding chapters, seven out of the ten patterns I've talked about are dry flies. This lack of balance does not indicate a personal preference for the floaters, however. Actually, during an average season I split my angling efforts almost equally between wets and dries; I think you catch more fish that way. I've also found that it's just harder to come up with wet-fly patterns that are truly innovative and still have something to offer. For one thing, wet flies have been around much longer than dries, and for another, the simplest, plainest patterns are often the best. As an example, take soft brown hackles, combine them with a peacock-herl body, and you have a hoary and unpretentious wet fly from the last century (or before) that's tough to beat.

And this thought brings me back to the Tadpole. Among the bright and shiny songs that colorful streamers sing, the Tadpole is something of a dirge. It reminds me of the bottom of an old coffee pot. I'm afraid the Tadpole may be too much a ragamuffin for traditionalists. Still, I couldn't bear to leave it out. Although it's not as well tested as the other patterns, I think it has real promise. As noted before, I'm not sure the trout necessarily take it for a minnow. I usually fish it that way in sizes 8 or 10, but the Tadpole has shown considerable ability tied on long-shank hooks all the way down to 16s, and this is why I suspect that it simulates nymphs and other food forms. (It's pretty rare to find an actual tadpole in a trout's gullet, so the fly's name is only meant to be descriptive.) And I really do have high hopes that the Tadpole may prove treacherous to larger trout. I promised some time ago not to lay any lunker lore on the reader, and I've been pretty good about it so far, but since we're nearing the end, I wonder if I might be permitted one whopper story as long as it's watered down.

One of the more charming streams I visit on a regular basis is the Chama River, a piece of fluid real estate shared between Colorado and New Mexico. The Chama's newborn east and west forks quite literally take high dives off the vertical face of imposing Banded Peak on the Colorado side of the line. Picking themselves up at the bottom, these rivulets are soothed by beaver-dammed springs before merging to flow due south into New Mexico. The main Chama reminds me of a bullfrog. It puffs up to an intimidating size during the runoff, when it's unfordable and almost unwadable, only to become a virtual pussycat of a river by August. This is noteworthy because the Chama has good enough supplies of food to support some very nice browns, and when the river does its Alice in Wonderland shrink trick in the late season, these trout get concentrated in scattered deeper pools. My log book from pre-Tadpole days shows that it took me an average of six or seven fishing days to catch just one big trout, and almost all of them were taken on a nymph or streamer rather than a dry fly. Gratifyingly, my new Tadpole is more than likely to seduce at least one and often several sizable browns on any given day, and this is the kind of performance that has encouraged me. I think back to the full-bodied 16- and 19-inch fish it took last year within a thirty-minute period from two pools a stone's throw apart. While the Chama holds larger trout, these were not fish that one would kick out of the creel, so to speak. (I did return them, though.)

After a whole series of semispectacular performances of this type in New Mexico and Colorado I felt the Tadpole had earned a shot at the big leagues of streamer fishing. Accordingly, in 1978 I took it on the road to Jackson Hole. It was my intention to match the Tadpole against the redoubtable Muddler Minnow in the Snake and its three major tributaries, the Hoback, Greys, and Gros Ventre Rivers. As it turned out, the trip was a near-disaster. The Tadpole's career almost came to an end on the very first morning when my somber fly absorbed a defeat so devastating that I seriously considered cashiering the pattern altogether. This unfortunate encounter took place on the broad and picturesque Gros Ventre, northeast of Jackson. Gary Troup, an old companion and experienced streamer fisherman, was to manipulate Jack Dennis's version of the Muddler, a fly of impeccable regional reputation. Meanwhile I would be at the Tadpole's controls. The plan was to leapfrog along, comparing notes as the day progressed, and it was about twenty-five minutes after we entered the river that our paths crossed for the first time. I had yet to see or feel a fish when Gary cut across a piece of rocky bank between meanders to join me. Approaching, he stumbled awkwardly as if excited or upset, and I thought he looked a bit dazed.

"Five," Gary blurted. "I've caught five already, all of them better than a foot." As proof he pulled a lovely 15-inch cutthroat from his creel, gill-hooked too deeply to save. This feat was further underscored by the fact that these were fat, tough fish, long on stamina and hooked in such heavy water it must have taken several minutes to land each of them. Thus fishing time between hookups had been minimal, and I knew that Gary had worked several small pools in the process just like the ones I had fished, so I was thoroughly convinced. I remember that it wasn't pleasant to replace my impotent Tadpole with a Muddler, but it was also no fun watching someone else haul in gorgeous trout, hand over fist, while I was drawing a blank.

Soon I began to catch some too, although Gary still continued to outnumber me by about three to one despite our identical Muddlers. Oddly enough, this turned out to be a stroke of luck. It was only because of the ongoing discrepancy in our catch rates that I noticed how different his method of presentation was from mine. I had been favoring a dead or natural drift, basically casting the muddler upstream and merely giving the fly an occasional twitch. By contrast, Gary, who had fished with area guides, used their local technique, casting either directly across currents or across and down. He would then immediately swing the streamer across the current briskly and perpendicular to the axis of flow until it hung below him, almost breaking surface. If the Muddler was still unattached to a cutthroat, he would strip-retrieve it via short rapid jerks, so the whole presentation was very active, with the fly barely sunken.

The practical crux of this observation amounted to a reprieve for the Tadpole. To summarize a good many hours of fishing, I found that if I presented *either* pattern in the prescribed manner, I did well (or the fly did). Indeed, it turned into a virtual draw between Muddler and Tadpole, and I was delighted to escape with a stalemate, too. When you're playing the champ on his home field, it's almost like winning.

Of course, I'd like to think that the Tadpole might even come out on top in certain situations. It really has much more intrinsic action than the Muddler, thanks to the marabou feather, and I hope to continue the contest. Perhaps the sculpin in the Jackson/West Yellowstone country are not so dark as the charcoal Tadpole, and it's interesting to speculate as to how the fly might perform dyed to an olive, brown,

yellow, or some combination thereof. The fun is the finding out!

In closing, I see the Tadpole as a dark, almost sinister form, globular and bullet-headed. It's like an oversized comma with a swinging tail, dancing and swimming through the water with some tumbles thrown in. How the trout see it, once again, I don't know. Nonetheless, this partially purloined patter appears to "turn them on," in the modern vernacular.

CHAPTER 12

Final Thoughts

When I first began to develop my own patterns some thirty years ago, I had a delusion to the effect that I was some sort of piscatorial Thomas Edison. Most of the fly fisherman I knew in Denver during the 1940s didn't tie themselves, and the few who did pretty much confined their efforts to old standards, patterns such as the Rio Grande King, Royal Coachman, Ginger Quill, and Black Gnat. It turns out that this was a delusion of grandeur on my part. If you browse through Mary Orvis Marbury's *Favorite Flies and Their Histories,* an anthology of comments by anglers from the Victorian era, it's clear that anglers everywhere were adding their own touches to established patterns and inventing brand-new ones. And if my gold-, pink-, and blue-winged patterns seem radical, just check the color plates in this very old book. (It was reprinted in 1955.) No doubt about it, fly fishers are creative by nature. If I'm an unusual tyer, it's probably only because of an inexplicable urge to write about my ventures at the vise. So with this prologue in mind, I'd like to offer a few words of advice based on personal trials and tribulations with the sincere hope that my remarks will not come across as condescending.

Let's start out with the indelible idea that the patterns you tie onto your leader have ever so much better a chance of catching fish than those that reside quietly in the darkness of a fly box. It also follows that the patterns *you* create are the ones you'll most want to try. These are the flies with which you'll fish longest and hardest. Here's a for instance: During the 1970 season I contracted an intense fascination for experimental streamer patterns tied with white marabou wings. I work with streamers a lot anyway, but that year I really outdid myself. The marabous saw action from March to November in waters of every kind from as far north as the Big Lost River in Idaho on the 44th Parallel to tiny Holy Ghost Pond in the desert foothills of central New Mexico on the 35th Parallel. I fished these streamers morning, noon, and night. In stubborn and probably foolish fashion, I pushed them at the trout even at times when other anglers were doing better with nymphs or dry flies. I remember wasting most of a magnificent mayfly hatch on Silver Creek this way, and when 1970 came to its end the marabous had taken nearly 300 trout or just over 50 percent of all the fish I caught that year, dry flies, nymphs, and other non-marabou streamers included. Did this prove that the white-winged marabou is a super pattern? In no way. At times it can be, but what this experience really proves is the old saw that the flies you fish with are the flies you'll catch fish with. To prove the point, by the time the next season rolled around I had new interests and used the marabou fly only when there was a "reason." As a consequence, in 1971, this streamer accounted for barely 2 percent of my total catch, about a dozen trout in all.

MARABOU MADNESS

As you can see, it's a setup. When a new pattern comes out of your vise, it's only human to want it to succeed, and it very probably will, at least to some extent. In this context I want to warn you especially about what might be called easy water, those natural fish hatcheries where eager customers can hardly wait for a fly to come along, natural or otherwise. In the Rockies there are many spots where Nature's larder is not that well stocked from the trout's point of view and fishing pressure is low. Thus an angler can reasonably expect an open-armed reception from the crowded inhabitants and can really clean up on the little trout if he wishes. It really amounts to a form of bully-ism, and you can get a pretty big head over the performance of one of your patterns in a situation such as this. As a rule it's far better to test one's tying triumphs in big-league waters if not in a variety of settings.

Still and all, thinking back, the remedial fishing holes can be instructive at times. Only 45 crow's-flight miles southwest of megalopolitan Denver there lies a still-rugged region flanked on one side by the refreshingly unimproved Platte and Kenosha ranges and on the other by the weathered but inaccessible Tarryall Mountains, the ancestral home of a splendid herd of bighorn sheep. In the valley between runs appropriately named Lost Park Creek. During high-school days it was a long and somewhat tense jeep trip, for segments of the old wagon road tilted so steeply it seemed that just one more degree of pitch would send the vehicle rolling into the water. Winding through a lengthy series of grassy meadows at an elevation of 10,000 feet, Lost Park Creek more resembled a famine-relief center than a trout stream. Although stark beginners couldn't catch more than half a dozen in an hour, for anyone else it was virtually a Salvelinus slaughter. We'd get bored and float all sorts of things down the pike just to watch the little fellows come up. Pine cones, apple cores, and plastic spinning bubbles always got a lot of attention; they'd bounce these objects around like so many trained dolphins, and yet I managed to invent a fly that rarely caught a thing. It wasn't any idle effort either; this was an innovative and (I thought) cleverly constructed water beetle replete with a two-tone body and six individually jointed legs, no less. I had acquired some plastic polymer which, when heated in my mother's double boiler, could be blended with pigment and molded around a hook. Before the stuff hardened, I'd insert six strands of copper wire from an electrical cord, trimmed to length and carefully bent into symmetrical joints. It was a crafty-looking imitation, if a bit contrived, but amazingly, the brookies reacted as if current were still running through the wire legs; they avoided my beetle like the plaque. Obviously, if small trout that are both ravenous and naive shun your pattern, it's likely a loser!

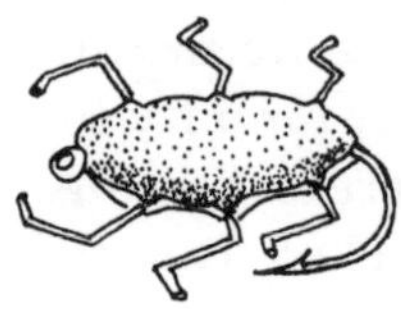

A beginners stream where the pickings are easy can also be quite helpful when conducting competitive matches between flies over a short period of time. I once became interested in studying changing ratios of brooks to cutthroat along the upper parts of Lime Creek, the major tributary of Colorado's Fryingpan River. Characteristically, when other species are introduced into a wilderness cutthroat stream, the red-gashed natives will progressively retreat to the highest, least-accessible portions of the system, and in the remote past some early settler or miner must have laboriously packed a few brookies past waterfalls into Lime Creek's headwaters. Here, as is their custom, these pilgrims proceeded to overpopulate their environment, thereby forcing the zero-population-growth (or at least less fecund) cutthroat ever higher. Following a glacial valley upward is much like climbing a staircase; there will be an alternating series of flats (like steps) and steep pitches or cliffs (risers). On Lime Creek I knew that at 10,500 feet the ratio of brooks to natives was approximately 19:1, but there was still a long series of glacial stairsteps to climb before reaching the cutthroat's last potential bastion, a meadow at the foot of the talus slopes spilling down from the very spine of the divide. Therefore, I expected that the ratio would eventually even out and possible reverse itself.

My plan was to identify about twenty trout at each level before climbing up to the next. I say "identify" because the fish were mostly small and the banks open, and with barbless hooks, I could dangle each one just long enough to see which it was before jiggling it off the hook, the quicker the better. I wanted a fly that would be plainly visible to both myself and the experimental subjects, so I decided on lightly weighted streamers, size 10, and selected a brown-hackled white-winged bucktail with a red wool body. This is a good old "meat and potatoes" pattern, and I had every intention of fishing it all day, for it seemed unlikely that they would be selective. I was disappointed. The bucktail produced just five in half an hour, and at that rate I knew I'd not have enough numbers for a meaningful result in one day. In an effort to get things going, I went to the Mickey

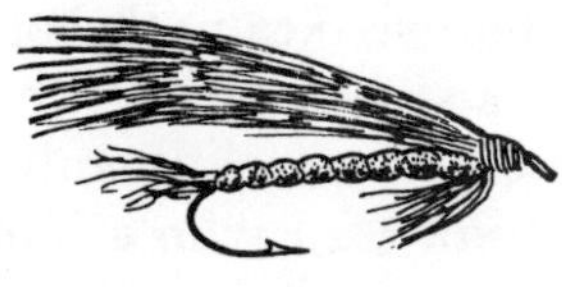

Finn, that gaudy gamin of the streamer family, only to elicit a strong vote of no confidence. The yellow-and-red streaker caught only two in twenty minutes. So I tried a Muddler Minnow. The Muddler was fairly new in those days, and I was relieved when it ran off a string of eight fish for me during the next half hour, but I tried one more pattern, a simple white-winged marabou with a peacock-herl body. It must have been the sensuous shimmy that's so special to this particular feather, because the bucktail also sported

a white wing and yet the marabou more than tripled the bucktail's take, an impressive eighteen in thirty minutes! At this turn of events, I became more fascinated by the struggle between streamers than with the ratio question, so I recycled through the four patterns, again beginning with the bucktail. This time it improved to seven in thirty minutes, while the Mickey Finn failed to score at all in fifteen minutes and got taken off the field. The Muddler then managed ten in half an hour, and the marabou once again came through to the tune of sixteen in half an hour.

Of course, it hadn't been my original intention to pit streamer patterns against one another. However, in the end I had a nice set of data that permitted a descriptive ranking of the four combatants (for that stream on that day) as follows:

1. Bucktail .... Fair
2. Mickey Finn .... Very poor
3. Muddler Minnow ... Good
4. Marabou .... Outstanding

The point is that one seldom has a chance to work with such huge numbers of helpers as on Lime Creek, and consequently it takes much longer to appreciate pattern preferences. You need a lot of numbers to see these differences clearly. Pattern A could be twice as effective as pattern B, averaging two trout per hour, but taking the luck factor into account and knowing you'll miss some solid strikes with both flies, it would likely take all day to prove A's superiority. Although shooting fish in rainbarrels may not be terribly challenging, rainbarrel trout will tell you what they think of a whole bunch of flies in just a few hours.

As an aside, I'd like to make a couple of remarks about "real fishing time" in the context of matches between patterns. There are just all sorts of ways that time can be lost on the stream. By this I mean those minutes during which a competing fly is not being actively presented. It might be roosting in a tree waiting to be liberated, jammed under a rock, or in the jaw of a sizable fish you're chasing downriver. And those seconds spent negotiating a slippery piece of river bottom, replacing a tippet, or fiddling with the fly add up, too. Over the long run these average out, but if each competitor gets only half an hour or so to show its stuff, they may not. For instance, quite unintentionally and through no fault of its own, pattern A could get snagged six times during its turn to just once for pattern B, with the result A loses fifteen of its thirty minutes. This is why I sometimes carry a stopwatch, ridiculous though it may sound. When a significant break in operations comes along, I hit the button, restarting when the interruption is over.

Don't try this if you're at all sensitive about being labeled as an oddball. I was standing in the Pecos River headwaters, east of Santa Fe, one evening making an entry in my log when two anglers surprised me. It so happened that I had a camera around my neck, a stopwatch pinned to one lapel, and a golfer's digital counter strapped to the other. (This is a wristwatch type of thing that counts up to 99 when you press the stem; for certain mathematical reasons, it didn't fit my game too well, so I use it for adding up strikes.) A small ruler, stream thermometer, flashlight, and surgical forceps (for hook extractions) protruded from various pockets, so I must have resembled a walking Science Fair project.

These fellows were obviously curious, and after I'd mumbled through a sort of explanation about the counter and stopwatch, one of them recommended a new pocket computer he'd seen. He pointed out that with something like this I could calculate standard deviations and progress to determinations of statistical probability from my data for programming and retrieval. As they moved off down the trail, I'm almost sure they were chuckling. I wonder if these weren't a pair of smart-aleck physicists from the nearby Los Alamos Laboratories. Anyhow, this kind of pseudo-research amuses me, and, after all, it is a harmless pastime.

I believe it's a very human mistake to become overly impressed with your tying creations. Parents of new patterns are like other parents; they have a way of wanting their offspring to do well and sometimes recognize excellence when there's really nothing special to report. I know I have an unfortunate habit of getting carried away with a construction idea and going into mass production before the concept in question has been shown to have true merit on the stream. The careful stone fly imitation in the chapter on the Pink Wolf is a case in point. And I remember other very pretty dry flies that proved to be thoroughgoing failures. Just imagine an olive-hackled pattern with a green quill body seductively set off by chartreuse wings, or a yellow-bodied floater with red-brown hackle artistically punctuated by hot-orange wings! (Embarrassingly, my wife salvaged these patterns, glued them to tiny hangers, and now wears them as earrings, perish the thought.) Then there was the ill-fated series of what I called "tractor-trailer" nymphs and streamers. I'd clip a hook shaft in two with wire cutters and epoxy a short strip of 4X monofilament to both ends, forming a hinge. When the body was tied, the hinge was left uncovered so that the rear end of the fly could flop around independent of the head. I put the first one together one snowy day in January, and it looked so good in the bathtub that by April I had dozens of tractor-trailers representing all sorts of patterns. So it was a disappointment to find that the trout could not have cared less whether a given fly was tied with or without swivel hips, and worse still, the things kept breaking at the hinge, so I was left with a collection of undependables.

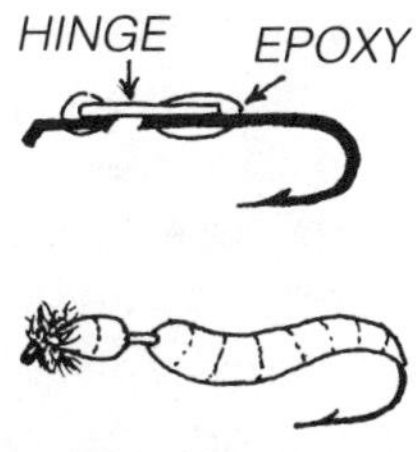

Probably the most imaginative failure I've perpetrated was an attempt to suggest the dragonfly nymph. We generally associate dragonflies with lakes or ponds. Nonetheless, there are several stream species, and these are potentially attractive creatures to copy in the nymph form, since they are large, meaty creatures. The nymphs are novel in that they propel themselves through the water by emitting a jet stream from their posteriors, rocket fashion, and it occured to me that this was an opportunity to achieve a new pinnacle of creative realism. I tied a segment of plastic drinking straw on top of and parallel to the hook shaft under the body materials, crimping the forward end slightly while leaving the rear end wide open so that a stream of bubbles issued from that orifice when the fly was cast into the water. Even the pattern's name was bad!

A friend recently startled me with what seemed a rather abrupt question. "Look," he said, "I've got just so much time, patience, and money for fooling with new flies; which of your crazy patterns is the best?" At the time I was a little taken aback, although I shouldn't have been. Certainly, these aren't patterns you'll find for sale; they have to be self-tied or specially ordered. And most of us don't have that many hours to spend on the water, nor can we try more than one fly at a time. Of course, his question was unanswerable, since it all depends on where, when, and even how you do your fishing. Fortunately, there is no such thing as a "can't fail" fly, so this is what I told him:

The Spanish Fly, Golden Adams, Laramie, and Maniac Nymph are clearly very dependable performers and not generally given to off days. Thus, they are staples in my fly boxes. The petite Little Hopper and simple Black Widow are staples, too, although I tend to reserve them for difficult quiet water and reluctant risers. These last two patterns are not so different from several well-known flies, but if the idea of showing the trout something that is going to be new and strange to them is appealing, I'd think about the Laramie, Pink Wolf, Blue Max, and Tadpole. And if unusual construction concepts strike your fancy, the Trashliners, Blue Max, and Tadpole should provide grist for one's mill. I hesitate to use the term "exciting" when talking about a small bit of steel, thread, fur, and feathers, although the Pink Wolf, Blue Max, and Tadpole qualify in this regard. Relatively newer than the rest of my flock, the early returns have been most encouraging, and I'm looking forward with considerable anticipation to learning more about these fanciful flies.

Taking the whole group of patterns together, they provide just lots of room for further experimentation and improvement, and that's precisely where the fun is. You don't really want to spend the rest of your angling days fishing with someone else's flies, do you?

Index

W

Y

Z

Spanish Fly

Golden Adams

Laramie

Maniac Nymph

Trashliner